Edward West Nichols

Analytic Geometry for Colleges, Universities and Technical Schools

Edward West Nichols

Analytic Geometry for Colleges, Universities and Technical Schools

ISBN/EAN: 9783744648905

Printed in Europe, USA, Canada, Australia, Japan

Cover: Foto ©berggeist007 / pixelio.de

More available books at **www.hansebooks.com**

ANALYTIC GEOMETRY

FOR

COLLEGES, UNIVERSITIES, AND TECHNICAL SCHOOLS.

BY

E. W. NICHOLS,

PROFESSOR OF MATHEMATICS IN THE VIRGINIA MILITARY INSTITUTE.

LEACH, SHEWELL, & SANBORN,
BOSTON. NEW YORK. CHICAGO.

PREFACE.

THIS text-book is designed for Colleges, Universities, and Technical Schools. The aim of the author has been to prepare a work for beginners, and at the same time to make it sufficiently comprehensive for the requirements of the usual undergraduate course. For the methods of development of the various principles he has drawn largely upon his experience in the class-room. In the preparation of the work all authors, home and foreign, whose works were available, have been freely consulted.

In the first few chapters *elementary examples* follow the discussion of each principle. In the subsequent chapters *sets of examples* appear at intervals throughout each chapter, and are so arranged as to partake both of the nature of a *review* and an *extension* of the preceding principles. At the end of each chapter *general examples*, involving a more extended application of the principles deduced, are placed for the benefit of those who may desire a higher course in the subject.

The author takes pleasure in calling attention to a "Discussion of Surfaces," by A. L. Nelson, M.A., Professor of Mathematics in Washington and Lee University, which appears as the final chapter in this work.

He takes pleasure also in acknowledging his indebtedness

to Prof. C. S. Venable, LL.D., University of Virginia, to Prof. William Cain, C.E., University of North Carolina, and to Prof. E. S. Crawley, B.S., University of Pennsylvania, for assistance rendered in reading and revising manuscript, and for valuable suggestions given.

                                        E. W. NICHOLS.

LEXINGTON, VA.

        January, 1893.

CONTENTS.

CONTENTS.

vii

CONTENTS.

CHAPTER IX.

GENERAL EQUATION OF THE SECOND DEGREE.

CHAPTER X.

HIGHER PLANE CURVES.

EQUATIONS OF THE THIRD DEGREE.

EQUATIONS OF THE FOURTH DEGREE.

TRANSCENDENTAL EQUATIONS.

PART II.— SOLID ANALYTIC GEOMETRY.

CHAPTER I.

CO-ORDINATES.

CHAPTER II.

THE PLANE.

CHAPTER III.

THE STRAIGHT LINE.

CONTENTS.

CHAPTER IV.

DISCUSSION OF SURFACES OF THE SECOND ORDER.

PLANE ANALYTIC GEOMETRY.

PART I.

CHAPTER I.

CO-ORDINATES.—THE CARTESIAN OR BILINEAR SYSTEM.

1. THE *relative* positions of objects are determined by referring them to some other objects whose positions are assumed as known. Thus we speak of Boston as situated in *latitude* — ° *north*, and *longitude* — ° *west*. Here the objects to which Boston is referred are the *equator* and the *meridian* passing through Greenwich. Or, we speak of Boston as being so many *miles north-east* of New York. Here the objects of reference are the *meridian of longitude* through New York and *New York itself.* In the first case it will be observed, Boston is referred to two lines which intersect each other at right angles, and the position of the city is *located* when we know its *distance* and *direction* from each of these lines.

In like manner, if we take any point such as P₁ (Fig. 1) in the plane of the paper, its position is fully determined when we know its *distance* and *direction* from each of the two lines O X and O Y which intersect each other at right angles in that plane. This method of locating points is known by the name of THE CARTESIAN, or BILINEAR SYSTEM. The lines of

1

reference O X, O Y, are called Co-ordinate Axes, and, when read separately, are distinguished as the X-axis and the Y-axis. The point O, the intersection of the co-ordinate axes, is called .the Origin of Co-ordinates, or simply the Origin.

The lines x' and y' which measure the distance of the point P_1 from the Y-axis and the X-axis respectively, are

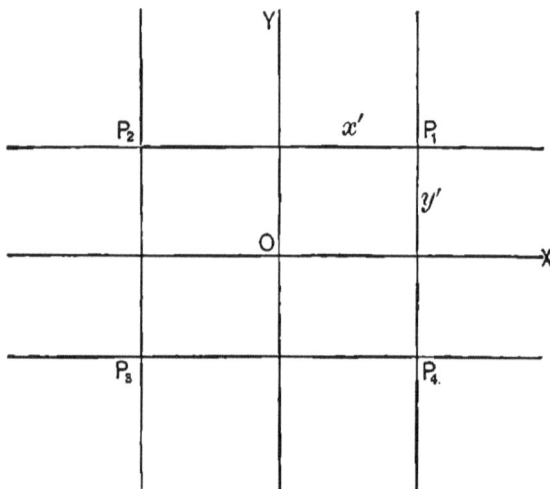

FIG. 1.

called the *co-ordinates of the point* — the distance (x') from the Y-axis being called the *abscissa* of the point, and the distance (y') from the X-axis being called the *ordinate* of the point.

2. Referring to Fig. 1, we see that there is a point in each of the four angles formed by the axes which would satisfy the conditions of being distant x' from the Y-axis and distant y' from the X-axis. This ambiguity vanishes when we combine the idea of direction with these distances. In the case of places on the earth's surface this difficulty is overcome by using the terms *north, south, east,* and *west.* In analytic geometry the algebraic symbols $+$ and $-$ are used to serve the same purpose. All distances measured to the *right* of the Y-axis are called *positive* abscissas; those measured to the left, *negative;* all distances measured *above* the X-axis are called *positive* ordinates; all distances *below, negative.* With this understanding, the co-ordinates of the point P_1 become (x', y'); of P_2, $(-x', y')$; of P_3, $(-x', -y')$; of P_4, $(x', -y')$.

3. The four angles which the co-ordinate axes make with each other are numbered 1, 2, 3, 4. The first angle is above the X-axis, and to the right of the Y-axis; the second angle is above the X-axis, and to the left of the Y-axis; the third angle is below the X-axis, and to the left of the Y-axis; the fourth angle is below the X-axis and to the right of the Y-axis.

EXAMPLES.

1. Locate the following points:

$(-1, 2)$, $(2, 3)$, $(3, -1)$, $(-1, -1)$, $(-2, 0)$, $(0, 1)$, $(0, 0)$, $(3, 0)$, $(0, -4)$.

2. Locate the triangle, the co-ordinates of whose vertices are,

$(0, 1)$, $(-1, -2)$, $(3, -4)$.

3. Locate the quadrilateral, the co-ordinates of whose vertices are,

$(2, 0)$, $(0, 3)$, $(-4, 0)$, $(0, -3)$.

What are the lengths of its sides?

Ans. $\sqrt{13}$, 5, 5, $\sqrt{13}$.

4. The ordinates of two points are each $= -b$; how is the line joining them situated with reference to the X-axis?

Ans. Parallel, below.

5. The common abscissa of two points is a; how is the line joining them situated?

6. In what angles are the abscissas of points positive? In what negative?

7. In what angles are the ordinates of points negative? In what angles positive?

4 *PLANE ANALYTIC GEOMETRY.*

8. In what angles do the co-ordinates of points have like signs ? In what angles unlike signs ?

9. The base of an equilateral triangle coincides with the X-axis and its vertex is on the Y-axis at the distance 3 below the origin ; required the co-ordinates of its vertices ?

Ans. $(\frac{1}{2} \sqrt{12}, 0)$, $(0, -3)$, $(-\frac{1}{2} \sqrt{12}, 0)$.

10. If a point so moves that the ratio of its abscissa to its ordinate is always $= 1$, what kind of a path will it describe, and how is it situated ?

Ans. A straight line passing through the origin, and making an angle of 45° with the X-axis.

11. The extremities of a line are the points $(2, 1)$, $(-1, -2)$: construct the line.

12. If the ordinate of a point is $= 0$, on which of the co-ordinate axes must it lie ? If the abscissa is $= 0$?

13. Construct the points $(-2, -3)$, $(2, 3)$, and show that the line joining them is bisected at $(0, 0)$.

14. Show that the point (m, n) is distant $\sqrt{m^2 + n^2}$ from the origin.

15. Find from similar triangles the co-ordinates of the middle point of the line joining $(2, 4)$, $(1, 1)$.

Ans. $(\frac{3}{2}, \frac{5}{2})$.

THE POLAR SYSTEM.

4. Instead of locating a point in a plane by referring it to two intersecting lines, we may adopt the second of the two methods indicated in Art. 1. The point P_1, Fig. 2, is fully determined when we know its *distance* O P_1 ($= r$) and *direction* P_1 O X ($= \theta$) from some given point O in some given line O X. If we give all values from 0 to ∞ to r, and all values from 0° to 360° to θ, it is easily seen that the position of every point in a plane may be located.

This method of locating a point is called the POLAR SYSTEM.

The point O is called the POLE; the line O X, the POLAR AXIS, or INITIAL LINE; the distance r, the RADIUS VECTOR; the angle θ, the DIRECTIONAL or VECTORIAL ANGLE. The distance r and the angle θ, (r, θ), are called the POLAR CO-ORDINATES of a point.

5. In measuring angles in this system, it is agreed (as in trigonometry), to give the positive sign $(+)$ to all angles meas-

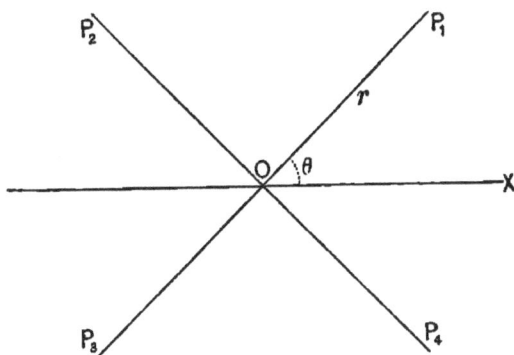

FIG. 2.

ured round to the left from the polar axis, and the opposite sign $(-)$ to those measured to the right. The radius vector (r) is considered as positive $(+)$ when measured from the pole *toward* the extremity of the arc (θ), and negative $(-)$ when measured from the pole *away from* the extremity of the arc (θ). A few examples will make this method of locating points clear.

If $r = 2$ inches and $\theta = 45°$, then $(2, 45°)$ locates a point P_1 2 inches from the pole, and on a line making an angle of $+45°$ with the initial line.

If $r = -2$ inches and $\theta = 45°$, then $(-2, 45°)$ locates a point P_3 2 inches from the pole, and on a line making an angle of 45° with the initial line also; but in this case the point is on that portion of the boundary line of the angle which has been produced *backward* through the pole.

If $r = 2$ inches and $\theta = -45°$, then $(2, -45°)$ locates a point P_4 2 inches from the pole, and out on a line lying below the initial line, and making an angle of 45° with it.

If $r = -2$ inches and $\theta = -45°$, then $(-2, -45°)$ locates a point P_2 directly opposite (with respect to the pole), the point P_4, $(2, -45°)$.

6. While the usual method in analytic geometry of expressing an angle is in *degrees, minutes,* and *seconds* (°, ′, ″), it frequently becomes convenient to express angles in terms of *the angle whose arc is equal in length to the radius of the measuring circle.* This angle is called the CIRCULAR UNIT.

We know from geometry that angles at the centre of the same circle are to each other as the arcs included between their sides; hence, if θ and θ' be two central angles, we have,

$$\frac{\theta}{\theta'} = \frac{arc}{arc'}.$$

Let $\theta' = $ unit angle; then $arc' = r$ (radius of measuring circle).

Hence $\qquad \dfrac{\theta}{circular\ unit} = \dfrac{arc}{r}.$

$$\therefore r\,\theta = arc \times circular\ unit.$$

If $\theta = 360°$, common measure, then $arc = 2\pi r$.

Hence, $\qquad r \times 360° = 2\pi r \times circular\ unit.$

Therefore the equation,

$$360° = 2\pi \times circular\ unit, \ \ \ . \ . \ . \ (1)$$

expresses the relationship between the two units of measure.

EXAMPLES.

1. What is the value in circular measure of an angle of 30°? From (1) Art. 6, we have,

$$360° = 30° \times 12 = 2\pi\ circular\ unit.$$

$$\therefore 30° = \frac{\pi}{6}\ circular\ unit.$$

2. What are the values in circular measure of the following angles?

1°, 45°, 60°, 90°, 120°, 180°, 225°, 270°, 360°.

3. What are the values in degrees of the following angles?

$$\frac{\pi}{3}, \frac{\pi}{2}, \frac{3}{4}\pi, \frac{3}{8}\pi, \frac{5}{4}\pi, \frac{1}{4}\pi, \frac{7}{8}\pi, \frac{5}{8}\pi, \frac{\pi}{6}, 2\pi.$$

4. What is the unit of circular measure?

Ans. 57°, 17′, 45″.

5. Locate the following points:

$(2, 40°)$, $\left(3, \frac{\pi}{2}\right)$, $(-4, 90)$, $(3, -135°)$, $(-1, -180°)$,

$\left(2, \frac{3}{4}\pi\right)$, $\left(1, -\frac{5}{4}\pi\right)$, $(-2, 270°)$, $(3, 2\pi)$,

$(-1, -\pi)$, $\left(2, -\frac{\pi}{2}\right)$, $\left(a, \frac{\pi}{4}\right)$.

6. Locate the triangle whose vertices are,

$\left(2, \frac{\pi}{8}\right)$, $\left(3, \frac{3}{4}\pi\right)$, $\left(1, \frac{5}{4}\pi\right)$.

7. The base of an equilateral triangle ($= a$) coincides with the initial line, and one of its vertices is at the pole; required the polar co-ordinates of the other two vertices.

Ans. $\left(a, \frac{\pi}{3}\right)$, $(a, 0)$.

8. The polar co-ordinates of a point are $\left(2, \frac{\pi}{4}\right)$. Give three other ways of locating the same point, using polar co-ordinates.

Ans. $\left(-2, \frac{5\pi}{4}\right)$, $\left(-2, -\frac{3\pi}{4}\right)$, $\left(2, -\frac{7\pi}{4}\right)$.

9. Construct the line the co-ordinates of whose extremities are $\left(2, \frac{\pi}{6}\right)$, $\left(-1, -\frac{\pi}{3}\right)$.

10. How is the line, the co-ordinates of two of its points being $\left(3, \dfrac{\pi}{4}\right)$, $\left(3, \dfrac{3\pi}{4}\right)$, situated with reference to the initial line?   *Ans.* Parallel.

Find the rectangular co-ordinates of the following points:

11. $\left(3, \dfrac{\pi}{3}\right).$ **13.** $\left(4, \dfrac{\pi}{2}\right).$

12. $\left(-3, \dfrac{\pi}{6}\right).$ **14.** $\left(-2, \dfrac{3}{4}\pi\right).$

CHAPTER II.

LOCI.

7. The Locus of an Equation *is the path described by its generatrix as it moves in obedience to the law expressed in the equation.*

The Equation of a Locus *is the algebraic expression of the law subject to which the generatrix moves in describing that locus.*

If we take any point P₃, equally distant from the X-axis and the Y-axis, and impose the condition that it shall so move

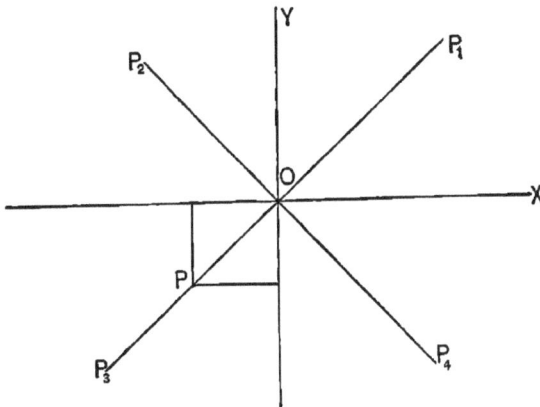

FIG. 3.

that the ratio of its ordinate to its abscissa shall always be equal to 1, it will evidently describe the line P₃P₁. The *algebraic expression* of this law is

$$\frac{y}{x} = 1, \text{ or } y = x,$$

and is called the *Equation of the Locus.*

The line P₃P₁ is called the *Locus of the Equation.* Again:

if we take the point P_4, equally distant from the axes, and make it so move that the ratio of its ordinate to its abscissa at *any* point of its path shall be equal to -1, it will describe the line $P_4 P_2$. In this case the equation of the locus is

$$\frac{y}{x} = -1, \text{ or } y = -x,$$

and the line $P_4 P_2$ is the locus of this equation.

8. It will be observed in either of the above cases (the first, for example), that while the point P_3 moves over the line $P_3 P_1$, its ordinate and abscissa while always equal are yet in a *constant state of change*, and pass through all values from $-\infty$, through 0, to $+\infty$. For this reason y and x are called the VARIABLE or GENERAL CO-ORDINATES of the line. If we consider the point at any particular position in its path, as at P, its co-ordinates $(-x', -y')$ are *constant* in value, and correspond to this position of the point, and to this position alone. The variable co-ordinates are represented by x and y, and the particular co-ordinates of the moving point for any definite position of its path by these letters with a dash or subscript; or by the first letters of the alphabet, or by numbers. Thus (x', y'), (x_1, y_1), (a, b), $(2, 2)$ correspond to some *particular* position of the moving point.

EXAMPLES.

1. Express in language the law of which $y = 3x + 2$ is the algebraic expression.

Ans. That a point shall so move in a plane that its ordinate shall always be equal to 3 times its abscissa plus 2.

2. A point so moves that its ordinate $+$ a quantity a is always equal to $\frac{1}{2}$ its abscissa $-$ a quantity b; required the algebraic expression of the law.

$$\text{\textit{Ans.} } y + a = \tfrac{1}{2}x - b.$$

3. The sum of the squares of the ordinate and abscissa of a moving point is always constant, and $= a^2$; what is the equation of its path?

$$\text{\textit{Ans.} } x^2 + y^2 = a^2.$$

4. Give in language the laws of which the following are the algebraic expressions:

$$2\,y = x - \frac{3}{2}.$$

$$x^2 - y^2 = -6.$$

$$\frac{x^2}{3} + \frac{y^2}{2} = 1.$$

$$xy = 16.$$

$$y^2 = 4\,x.$$

$$4\,x^2 - 5\,y^2 = -18.$$

$$2\,x^2 + 3\,y^2 = 6.$$

$$a^2y^2 + b^2x^2 = a^2b^2.$$

$$y^2 = 2\,px.$$

9. As the relationship between a locus and its equation constitutes the fundamental conception of Analytic Geometry, it is important that it should be clearly understood before entering upon the treatment of the subject proper. We have been accustomed in algebra to treat every equation of the form $y = x$ as indeterminate. Here we have found that this equation admits of a *geometric interpretation ;* i.e., that it represents a straight line passing through the origin of co-ordinates and making an angle of 45° with the X-axis. We shall find, as we proceed, that *every equation*, algebraic or transcendental, which does not involve more than *three* variable quantities, is susceptible of a *geometric interpretation.* We shall find, conversely, that *geometric forms* can be expressed *algebraically*, and that all the properties of these forms may be deduced from their algebraic equivalents.

Let us now assume the equations of several loci, and let us locate and discuss the geometric forms which they represent.

10. *Locate the geometric figure whose algebraic equivalent is*

$$y = 3\,x + 2.$$

We know that the point where this locus cuts the Y-axis has its abscissa $x = 0$. If, therefore, we make $x = 0$ in the equation, we shall find the ordinate of this point. Making the substitution we find $y = 2$. Similarly, the point where the locus cuts the X-axis has 0 for the value of its ordinate. Making $y = 0$ in the equation, we find $x = -\frac{2}{3}$. Drawing now the axes and marking on them the points

$$(0, 2), \left(-\frac{2}{3}, 0\right),$$

we will have two points of the required locus. Now make x successively equal to

$$1, 2, 3, -1, -2, -3, \text{etc.}$$

in the equation, and find the corresponding values of y. For convenience let us tabulate the result thus:

Values of x	Corresponding	Values of y
1	"	5
2	"	8
3	"	11
− 1	"	− 1
− 2	"	− 4

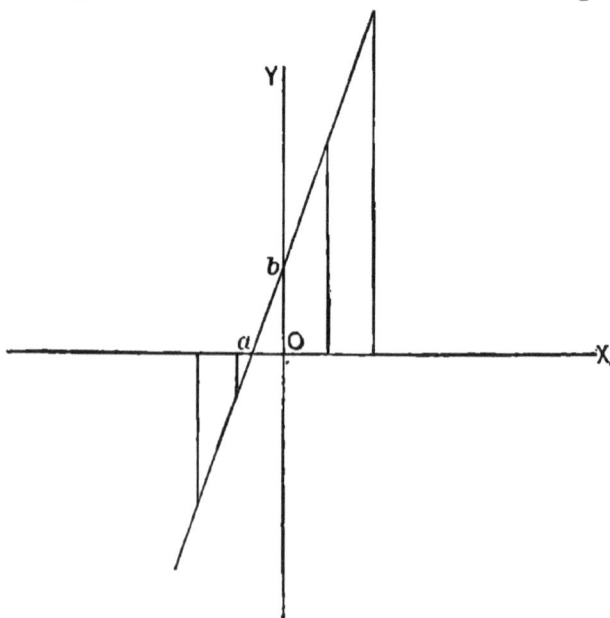

Fig. 4.

Locating these points and tracing a line through them we have the required locus. This locus *appears* to be a straight line — and it is, as we shall see hereafter. We shall see also that *every* equation of the *first degree* between two variables represents some straight line. The distances Oa and Ob which the line cuts off on the co-ordinate axes are called INTERCEPTS. In locating *straight* lines it is usually sufficient to determine these distances, as the line drawn through their extremities will be the locus of the equation from which their values were obtained.

EXAMPLES.

1. Locate the geometric equivalent of

$$\frac{1}{2} y - x = 1 - 2 x.$$

Solving with respect to y in order to simplify, we have,

$$y = - 2 x + 2.$$

The extremities of the intercepts are

$$(0, 2), \quad (1, 0).$$

Locating these points, and drawing a straight line through them, we have the required locus.

Construct the loci of the following equations:

2. $y = - 2 x - 2.$

3. $y = 3 x - 1.$

4. $y = ax + b.$

5. $\frac{1}{2} y = cx - d.$

6. $2 y = 3 x.$

7. $\frac{y}{4} + 2 x = 3 x - y.$

8. $2 x + 3 y = 7 - y.$

9. $\frac{x - 1}{2} = \frac{y - 2}{3}.$

10. $1 - \frac{y - 2}{2} + x = \frac{2 x - 2}{3} + y.$

11. $x - y = - \frac{3}{2} y - 2 x.$

12. Is the point $(2, 1)$ on the line whose equation is $y = 2 x - 3$? Is $(6, 9)$? Is $(5, 4)$? Is $(0, - 3)$?

NOTE. — If a point is on a line, its co-ordinates must satisfy the equation of the line.

13. Which of the following points are on the locus of the equation $3x^2 + 2y^2 = 6$?

$$(2, 1), \; (\sqrt{2}, 0), \; (0, \sqrt{3}), \; (-1, 3), \; (-\sqrt{2}, 0), \; (2, \sqrt{3})$$

14. Write six points which are on the line

$$\frac{1}{2}y - 2x = 3y - 6.$$

15. Construct the polygon, the equations of whose sides are

$$y = -2x - 1, \; y = x, \; y = 5.$$

16. Construct the lines $y = sx + b$ and $y = sx + 4$, and show by similar triangles that they are parallel.

11. *Discuss and construct the equation:*

$$x^2 + y^2 = 16.$$

Solving with respect to y, we have,

$$y = \pm\sqrt{16 - x^2}.$$

The double sign before the radical shows us that for every value we assume for x there will be two values for y, equal and with contrary signs. This is equivalent to saying that for every point the locus has *above* the X-axis there is a corresponding point *below* that axis. Hence the locus is *symmetrical* with respect to the X-axis. Had we solved the equation with respect to x a similar course of reasoning would have shown us that the locus is also symmetrical with respect to the Y-axis. Looking under the radical we see that any value of x less than 4 (positive or negative) will always give two *real* values for y; that $x = \pm 4$ will give $y = \pm 0$, and that any value of x greater than ± 4 will give imaginary values for y. Hence the locus does not extend to the right of the Y-axis farther than $x = +4$, nor to the left farther than $x = -4$.

Making $\qquad x = 0$, we have $y = \pm 4$
" $\qquad y = 0$, " " $x = \pm 4$.

Drawing the axes and constructing the points,
(0, 4), (0, — 4), (4, 0), (— 4, 0), we have four points of
the locus; i.e., B, B′, A, A′.

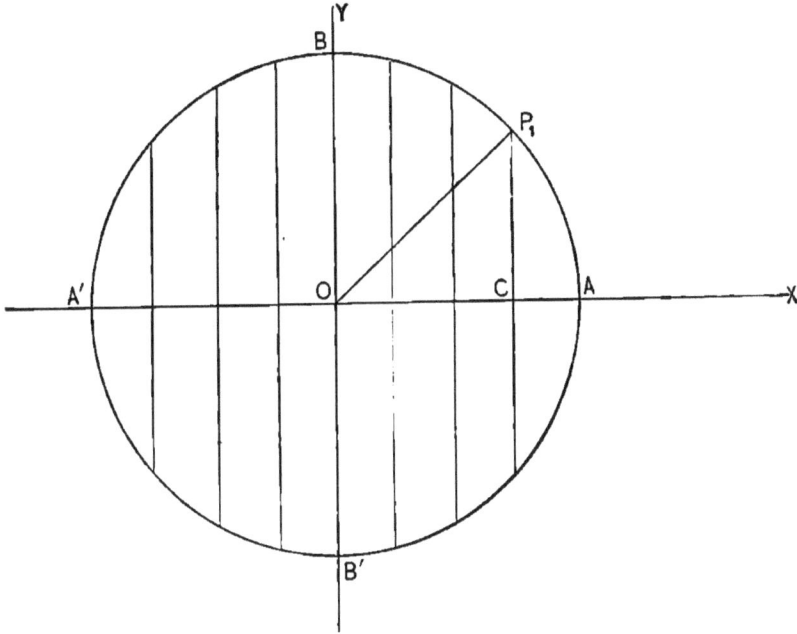

FIG. 5.

Values of x	Corresponding	Values of y
1	"	$+ 3.8$ and $- 3.8$
2	"	$+ 3.4$ and $- 3.4$
3	"	$+ 2.6$ and $- 2.6$
4	"	± 0
— 1	"	$+ 3.8$ and $- 3.8$
— 2	"	$+ 3.4$ and $- 3.4$
— 3	"	$+ 2.6$ and $- 2.6$
— 4	"	± 0

Constructing these points and tracing the curve, we find it
to be a circle.

This might readily have been inferred from the form
of the equation, for we know that the sum of the squares

of the abscissa (OC) and ordinate (CP$_1$) of any point P$_1$ in the circle is equal to the square of the radius (OP$_1$). We might, therefore, have constructed the locus by taking the origin as centre, and describing a circle with 4 as a radius.

Note. $x = \pm 0$ for any assumed value of y, or $y = \pm 0$, for any assumed value of x always indicates a *tangency*. Referring to the figure we see that as x increases the values of y decrease and become ± 0 when $x = 4$. Drawing the line represented by the equation $x = 4$, we find that it is tangent to the curve. We shall see also as we proceed that any *two coincident values* of either variable arising from an assumed or given value of the other indicates a point of tangency.

12. Construct and discuss the equation

$$9x^2 + 16y^2 = 144.$$

Solving with respect to y, we have

$$y = \pm \sqrt{\frac{144 - 9x^2}{16}}.$$

$$x = 0 \text{ gives } y = \pm 3;$$
$$y = 0 \quad \text{``} \quad x = \pm 4.$$

Drawing the axes and laying off these distances, we have four points of the locus; i.e., B, B', A, A'. Fig. 6.

Values of x	Corresponding	Values of y
1	"	$+2.9$ and -2.9
2	"	$+2.6$ " -2.6
3	"	$+2$ " -2
4	"	± 0
-1	"	$+2.9$ " -2.9
-2	"	$+2.6$ " -2.6
-3	"	$+2$ " -2
-4	"	± 0

Locating these points and tracing the curve through them, we have the required locus. Referring to the value of y we see from the double sign that the curve is symmetrical with respect to the X-axis. The form of the equation (containing only the second powers of the variables), shows that the locus is also symmetrical with respect to the Y-axis. Looking

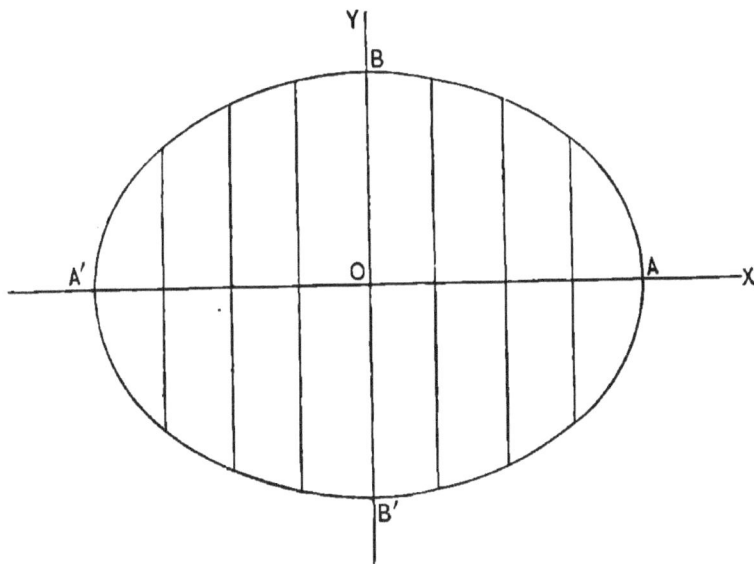

FIG. 6.

under the radical we see that any value of x between the limits $+4$ and -4 will give two real values for y; and that any value beyond these limits will give imaginary values for y. Hence the locus is entirely included between these limits.

This curve, with which we shall have more to do hereafter, is called the ELLIPSE.

13. *Discuss and construct the equation*

$$y^2 = 4 x.$$

Solving, we have

$$y = \pm \sqrt{4 x}.$$

We see that the locus is symmetrical with respect to the X-axis, and as the equation contains only the first power of x, that it is *not* symmetrical with respect to the Y-axis. As *every* positive value of x will always give real values for y, the locus must extend infinitely in the direction of the positive abscissæ; and as any negative value of x will render y

imaginary, the curve can have no point to the left of the
Y-axis.   Making $x = 0$, we find $y = \pm\,0$; hence the curve
passes through the origin, and is tangent to the Y-axis.
Making $y = 0$, we find $x = 0$; hence the curve cuts the
X-axis at the origin.

Values of x	Corresponding	Values of y
1	"	$+ 2$ and $- 2$
2	"	$+ 2.8$ " $- 2.8$
3	"	$+ 3.4$ " $- 3.4$
4	"	$+ 4$ " $- 4$

From these data we easily see that the locus of the equation
is represented by the figure below.

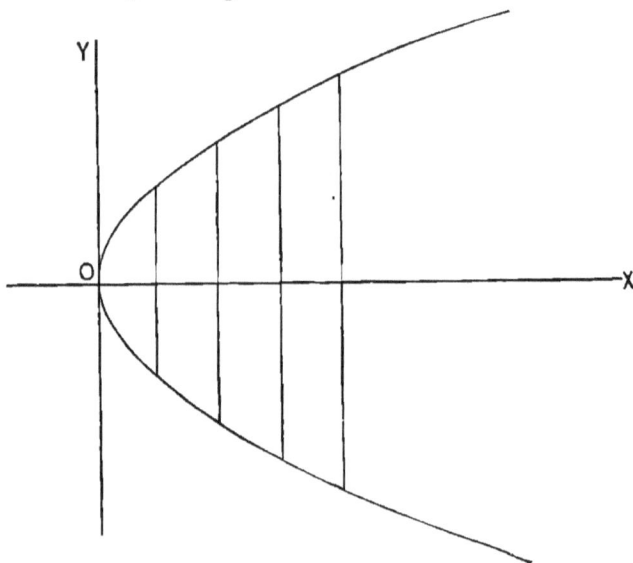

FIG. 7.

This curve is called the PARABOLA.

14. *Discuss and construct the equation*
$$4\,x^2 - 9\,y^2 = 36.$$
Hence
$$y = \pm\sqrt{\frac{4\,x^2 - 36}{9}}.$$

We see from the form of the equation that the locus must
be symmetrical with respect to both axes.   Looking under

the radical, we see that any value of x numerically *less* than
$+3$ or -3 will render y imaginary. Hence there is no
point of the locus within these limits. We see also that
any value of x greater than $+3$ or -3 will always give real
values for y. The locus therefore extends infinitely in the
direction of both the positive and negative abscissæ from the
limits $x = \pm 3$.

Making $x = 0$, we find $y = \pm 2 \sqrt{-1}$; hence, the curve·
does not cut the Y-axis.

Making $y = 0$, we find $x = \pm 3$; hence, the curve cuts
the X-axis in two points $(3, 0)$, $(-3, 0)$.

Value of x.	Corresponding.	Values of y
4	"	$+1.7$ and -1.7
5	"	$+2.6$ " -2.6
6	"	$+3.4$ " -3.4
-4	"	$+1.7$ " -1.7
-5	"	$+2.6$ " -2.6
-6	"	$+3.4$ " -3.4

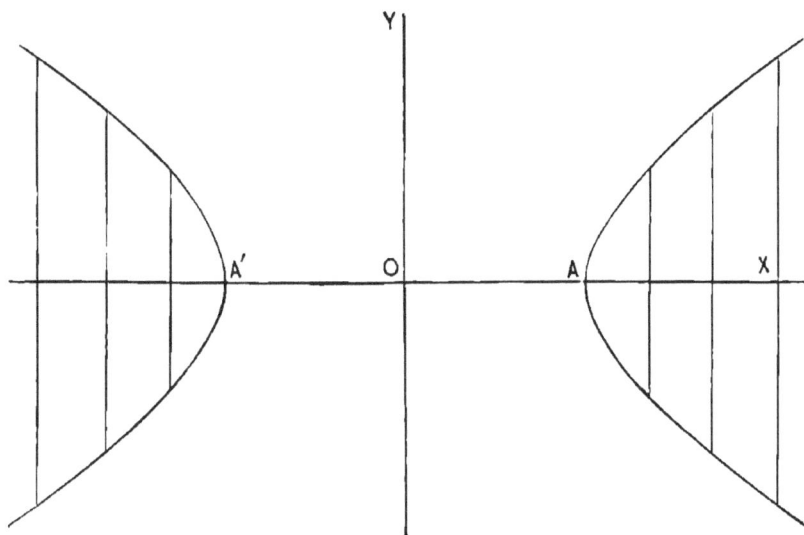

FIG. 8.

This curve is called the HYPERBOLA.

15. We have in the preceding examples confined ourselves to the construction of the loci of Rᴇᴄᴛᴀɴɢᴜʟᴀʀ equations; i.e., of equations whose loci were referred to rectangular axes. Let us now assume the Pᴏʟᴀʀ equation

$$r = 6 \, (1 - \cos \theta)$$

and discuss and construct it.

Assuming values for θ, we find their cosines from some convenient table of Natural Cosines. Substituting these values, we find the corresponding values of r.

Values of θ	Values of cos θ	Values of r
0	1.	$6 \, (1 - 1 \ ) = \quad 0$
30°	.86	$6 \, (1 - .86) = \quad .84$
60°	.50	$6 \, (1 - .50) = \quad 3.$
90°	0	$6 \, (1 - 0 \ ) = \quad 6.$
120°	$-.50$	$6 \, (1 + .50) = \quad 9.$
160°	$-.94$	$6 \, (1 + .94) = 11.64$
180°	$-1.$	$6 \, (1 + 1 \ ) = 12.$
200°	$-.94$	$6 \, (1 + .94) = 11.64$
240°	$-.50$	$6 \, (1 + .50) = \quad 9.$
270°	0	$6 \, (1 - 0 \ ) = \quad 6.$
300°	.50	$6 \, (1 - 50) = \quad 3.$
330°	.86	$6 \, (1 - .86) = \quad .84$

Draw the initial line OX, and assume any point O as the pole. Through this point draw a series of lines, making the assumed angles with the line OX, and lay off on them the corresponding values of r. Through these points, tracing a smooth curve, we have the required locus.

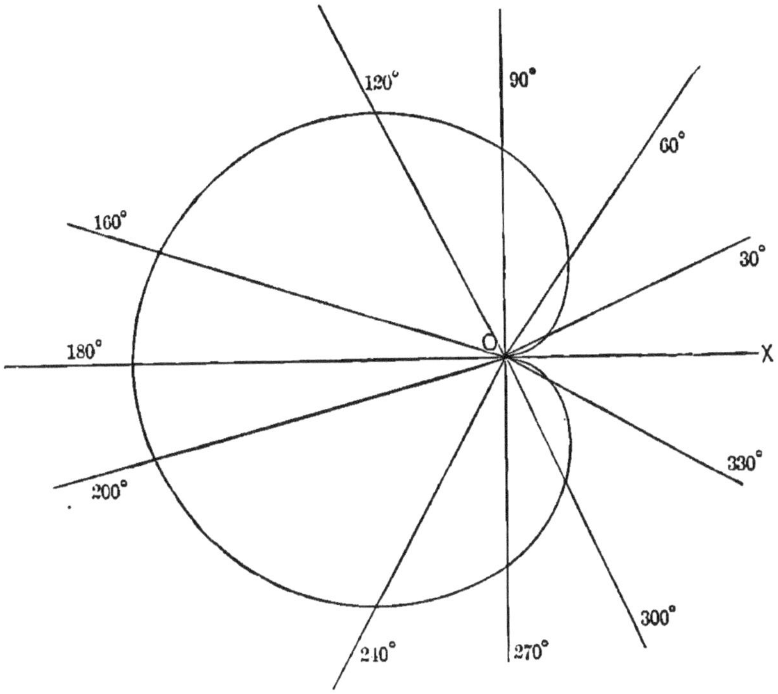

FIG. 9.

This curve, from its heart-like shape, is called the CARDIOID.

16. *Discuss and construct the transcendental equation*

$$y = \log x.$$

NOTE. — A transcendental equation is one whose degree *transcends* the power of analysis to express.

Passing to equivalent numbers we have $2^y = x$, when 2 is the base of the system of logarithms selected.

As the base of a system of logarithms can never be negative, we see from the equation that no negative value of x can satisfy it. Hence the locus has none of its points to the left of the Y-axis. On the other hand, as every positive value of x will give real values for y, we see that the curve extends infinitely in the direction of the positive abscissæ.

If $y = 0$, then

$$2^0 = x \therefore 0 = \log x \therefore x = 1.$$

If $x = 0$, then

$$2^y = 0 \therefore y = \log 0 \therefore y = -\infty.$$

The locus, therefore, cuts the X-axis at a unit's distance on the positive side, and continually approaches the Y-axis without ever meeting it. It is further evident that whatever be the base of the system of logarithms, these conditions must hold true for all loci whose equations are of the form $a^y = x$.

Values of x	Corresponding	Values of y
1	"	0
2	"	1
4	"	2
8	"	3
.5	"	-1
.25	"	-2

Locating these points, the curve traced through them wiil be the required locus.

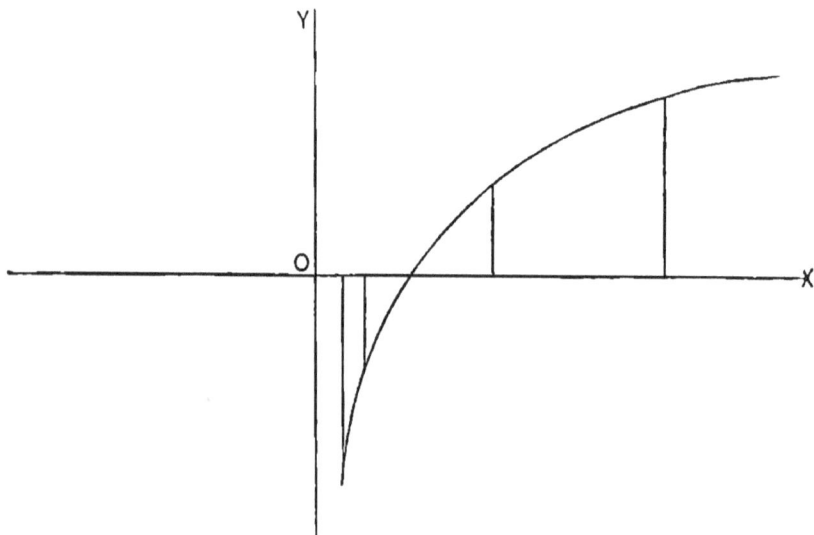

FIG. 10.

This curve is called the LOGARITHMIC Curve, its name being taken from its equation.

17. The preceding examples explain the method employed in constructing the locus of any equation. While it is true that this method is at best approximate, yet it may be made sufficiently accurate for all practical purposes by assuming for one of the variables values which differ from each other by very small quantities. It frequently happens (as in the case of the circle) that we may employ other methods which are entirely accurate.

18. In the discussion of an equation the first step, usually, is to solve it with respect to one of the variables which enter into it. The question of which variable to select is immaterial in principle, yet considerations of simplicity and convenience render it often times of great importance. The sole difficulty, in the discussion of almost all the higher forms of equations, consists in resolving them. If this difficulty can be overcome, there will be no trouble in tracing the locus and discussing it. If, as frequently happens, no trouble arises in the solution of the equation with respect to one of the variables, then that one should be selected as the dependent variable, and its value found in terms of the other. If it is equally convenient to solve the equation with respect to either of the variables which enter into it, then that one should be selected whose value on inspection will afford the simpler discussion.

EXAMPLES.

Construct the loci of the following equations:

1. $2y - 4x + 1 = 0.$
2. $y^2 - x^2 = 16.$
3. $2y^2 + 5x^2 = 10.$
4. $4x^2 - 9y^2 = -36.$
5. $y^2 + 4x = 0.$
6. $x^2 + y^2 - 25 = 0.$
7. $r^2 = a^2 \cos 2\theta.$
8. $x = \log y.$

Construct the loci of the following :

9. $x^2 - y^2 = 0.$

10. $x^2 + 2\,ax + a^2 = 0.$

11. $x^2 - a^2 = 0.$

12. $y^2 - 9 = 0.$

13. $y^2 - 2\,xy + x^2 = 0.$

14. $x^2 - x - 6 = 0.$

15. $x^2 + x - 6 = 0.$

16. $x^2 + 4\,x - 5 = 0.$

17. $x^2 - 7\,x + 12 = 0.$

18. $x^2 + 7\,x + 10 = 0.$

NOTE. — Factor the first member : equate each factor to 0, and construct separately.

CHAPTER III.

THE STRAIGHT LINE.

19. *To find the equation of a straight line, given the angle which the line makes with the* X*-axis, and its intercept on the* Y*-axis.*

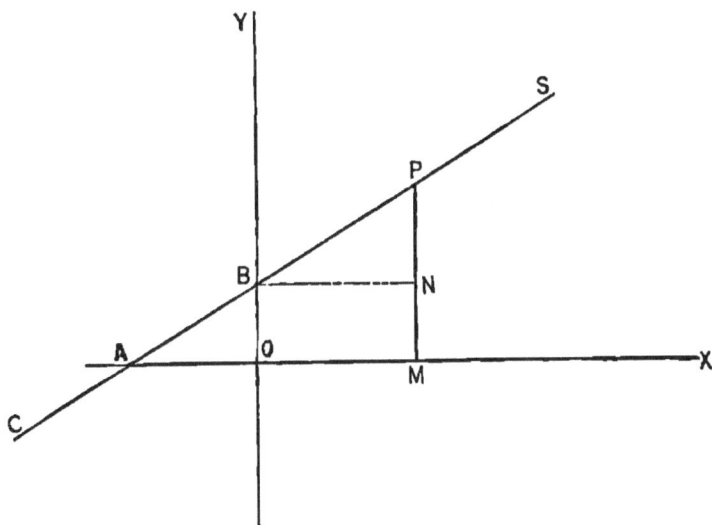

Fig. 11.

Let C S be the line whose equation we wish to determine. Let SAX $= \alpha$ and OB $= b$. Take *any* point P on the line and draw PM ∥ to OY and BN ∥ to OX.

Then (OM, MP) $= (x, y)$ are the co-ordinates of P.

From the figure PM = PN + OB = BN tan PBN $+ b$, but BN = OM $= x$, and tan PBN = tan SAX = tan α.

∴ Substituting and letting tan $\alpha = s$, we have,

$$y = sx + b \ldots (1)$$

Since equation (1) is true for *any* point of the line SC, it is true for *every* point of that line; hence it is the equation of the line. Equation (1) is called the SLOPE EQUATION OF THE STRAIGHT LINE; s (= tan α) is called the *slope*.

COROLLARY 1. If $b = 0$ in (1), we have,
$$y = s\,x \;\ldots\; (2)$$
for the slope equation of a line which passes through the origin.

COR. 2. If $s = 0$ in (1), we have
$$y = b$$
which is, as it ought to be, the equation of a line parallel to the X-axis.

COR. 3. If $s = \infty$, then $\alpha = 90°$, and the line becomes parallel to the Y-axis.

Let the student show by an independent process that the equation of the line will be of the form $x = a$.

SCHOLIUM. We have represented by α the angle which the line makes with the X-axis. As this angle may be either *acute* or *obtuse*, s, its tangent, may be either *positive* or *negative*. The line may also cut the Y-axis either *above* or *below* the origin; hence, b, its Y-intercept, may be either *positive* or *negative*. From these considerations it appears that
$$y = -\,sx + b$$
represents a line crossing the first angle;
$$y = sx + b$$
represents a line crossing the second angle;
$$y = -\,sx - b$$
represents a line crossing the third angle;
$$y = sx - b$$
represents a line crossing the fourth angle.

EXAMPLES.

1. The equation of a line is $2y + x = 3$; required its slope and intercepts.

Solving with respect to y, we have,

$$y = -\frac{1}{2}x + \frac{3}{2}.$$

Comparing with (1) Art. 19, we find $s = -\frac{1}{2}$ and $b = \frac{3}{2}$ = Y-intercept. Making $y = 0$ in the equation, we have $x = 3$ = X-intercept.

2. Construct the line $2y + x = 3$.

The points in which the line cuts the axes are

$$\left(0, \frac{3}{2}\right), \text{ and } (3, 0).$$

Laying these points off on the axes, and tracing a straight line through them, we have the required locus. Or otherwise thus: solving the equation with respect to y, we have,

$$y = -\frac{1}{2}x + \frac{3}{2}.$$

Lay off $OB = b = \frac{3}{2}$; draw $BN \parallel OX$ and make it $= 2$, also $NP \parallel OY$ and make it $= +1$. The line through P and B is the required locus.

For $\dfrac{PN}{NB} = \dfrac{1}{2} = \tan PBN$ $= - \tan BAX$.

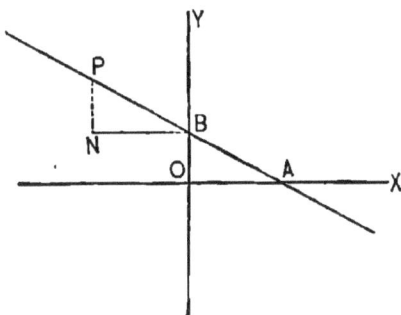

$$\therefore \tan BAX = s = -\frac{1}{2}.$$

3. Construct the line $2\,y - x = 3$.

Solving with respect to y, we have,

$$y = \frac{1}{2}\,x + \frac{3}{2}\,.$$

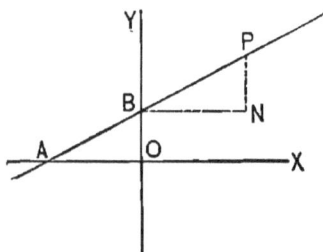

Lay off $BO = b = \dfrac{3}{2}$. Draw BN $\parallel$ to OX and make it $= 2$; draw also NP $\parallel$ to OY and make it $= 1$. A straight line through P and B will be the required locus.

For $\dfrac{\text{PN}}{\text{NB}} = \dfrac{1}{2} = \tan \text{ PBN} = \tan$

BAX $= s$.   Hence, in general, BN is laid off to *the right* or to *the left* of Y according as the coefficient of x is *positive* or *negative*.

Give the slope and intercepts of each of the following lines and construct:

4. $2\,y + 3\,x - 2 = 0$.

Ans. $s = -\dfrac{3}{2},\ b = 1,\ a = \dfrac{2}{3}\,.$

5. $x - 2\,y + 3 = 0$.

Ans. $s = \dfrac{1}{2},\ b = \dfrac{3}{2},\ a = -\,3.$

6. $6\,x + \dfrac{1}{2}\,y + 1 = 0$.

Ans. $s = -\,12,\ b = -\,2,\ a = -\dfrac{1}{6}\,.$

7. $\dfrac{x-2}{3} + \dfrac{y-2}{2} = 4$.   **8.** $\dfrac{y-1}{3} + 2\,x = 1 - y$.

9. $x + 2 + \dfrac{y}{2} = 4$.

Note. — a and b in the answers above denote the X-intercept and the Y-intercept, respectively.

What angle does each of the following lines cross ?

10. $y = 3x + 1.$ **12.** $y = 2x - 1.$

11. $y = -x + 2.$ **13.** $y = -3x - 2.$

14. Construct the figure the equation of whose sides are
$$2y + x - 1 = 0,\ 3y = 2x + 2,\ y = -x - 1.$$

15. Construct the quadrilateral the equations of whose sides are
$$x = 3,\ y = -x + 1,\ y = 2,\ x = 0.$$

20. *To find the equation of a straight line in terms of its intercepts.*

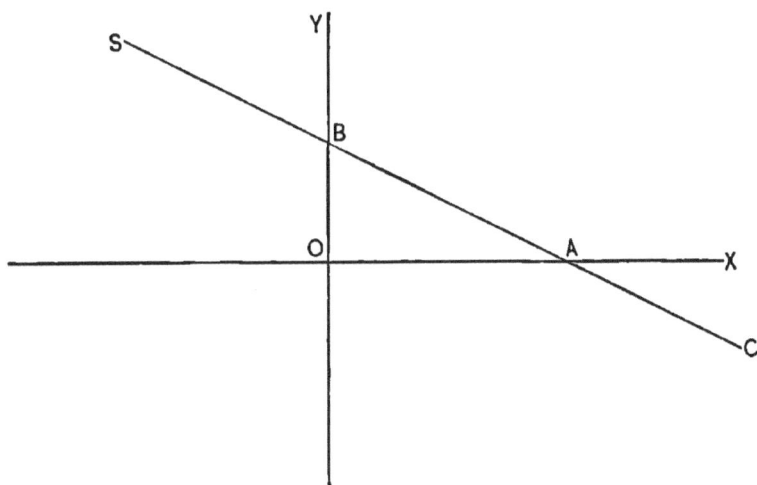

Fig. 12.

Let S C be the line.

Then $OB = b = $ Y-intercept, and
$OA = a = $ X-intercept.

The slope equation of a line we have determined to be Art. 19, equation (1),
$$y = sx + b.$$

From the right angled triangle AOB, we have,

$$\tan OAB = -\tan BAX = -s = \frac{OB}{OA}.$$

$$\therefore s = -\frac{b}{a}.$$

Substituting in the slope equation, we have,

$$y = -\frac{b}{a}x + b;$$

$$\therefore \frac{x}{a} + \frac{y}{b} = 1 \ldots (1)$$

This is called the SYMMETRICAL EQUATION of the straight line.

COR. 1. If $a +$ and $b +$, then we have,

$$\frac{x}{a} + \frac{y}{b} = 1, \text{ for a line crossing the first angle.}$$

If $a -$ and $b +$, then

$$-\frac{x}{a} + \frac{y}{b} = 1 \text{ is a line crossing the second angle.}$$

If $a -$ and $b -$, then

$$-\frac{x}{a} - \frac{y}{b} = 1 \text{ is a line crossing the third angle.}$$

If $a +$ and $b -$, then

$$\frac{x}{a} - \frac{y}{b} = 1 \text{ is a line crossing the fourth angle.}$$

EXAMPLES.

1. Construct $\frac{x}{3} - \frac{y}{2} = 1$.

NOTE. — Lay off 3 units on the X-axis and -2 units on the Y-axis. Join their extremities by a straight line.

Across which angles do the following lines pass ?

Give the intercepts of each, and construct.

2. $\dfrac{x}{3} + \dfrac{y}{2} = 1.$ **4.** $-\dfrac{x}{2} - \dfrac{y}{4} = 1.$

3. $-\dfrac{x}{3} + \dfrac{y}{3} = 1.$ **5.** $\dfrac{x}{5} - \dfrac{y}{7} = 1.$

Write the slope equations of the following lines, and construct:

6. $\dfrac{x}{5} - \dfrac{y}{6} = 1.$

$Ans.\ y = \dfrac{6}{5}x - 6.$

7. $\dfrac{x}{3} - \dfrac{y}{7} = 1.$

$Ans.\ y = \dfrac{7x}{3} - 7.$

8. $\dfrac{y}{2} + \dfrac{x}{6} = -1.$

$Ans.\ y = -\dfrac{1}{3}x - 2.$

9. $-y - \dfrac{x}{5} = 1.$

$Ans.\ y = -\dfrac{1}{5}x - 1.$

10. Write $y = sx + b$ in a symmetrical form.

$Ans.\ \dfrac{y}{b} - \dfrac{x}{\frac{b}{s}} = 1.$

Given the following equations of straight lines, to write their slope and symmetrical forms:

11. $2y + 3x - 7 = x + 2.$ **13.** $\dfrac{y-2}{x} = 3.$

12. $\dfrac{y-1}{2} = \dfrac{x-3}{3}.$ **14.** $\dfrac{x-y}{4} = \dfrac{2x-1}{3}.$

21. *To find the equation of a straight line in terms of the perpendicular to it from the origin and the directional cosines of the perpendicular.*

NOTE. — The Directional cosines of a line are the cosines of the angles which it makes with the co-ordinate axes.

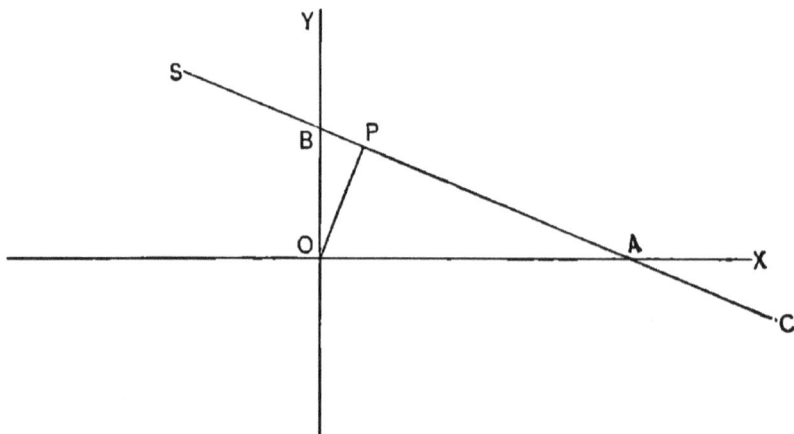

FIG. 13.

Let CS be the line.

Let OP $= p$, BOP $= \gamma$, AOP $= \alpha$.

From the triangles AOP and BOP, we have

$$OA = \frac{OP}{\cos \alpha}, \quad OB = \frac{OP}{\cos \gamma};$$

that is, $a = \dfrac{p}{\cos \alpha}, b = \dfrac{p}{\cos \gamma}.$

Substituting these values in the symmetrical equation,

Art. 20. (1), $\frac{x}{a} + \frac{y}{b} = 1$, we have, after reducing,

$$x \cos \alpha + y \cos \gamma = p \ \ \ldots \ (1)$$

which is the required equation.

Since $\gamma = 90° - \alpha$, $\cos \gamma = \sin \alpha$; hence

$$x \cos \alpha + y \sin \alpha = p \;\; \cdots \;\; (2)$$

This form is more frequently met with *than that given in* (1) and is called the NORMAL EQUATION of the straight line.

Cor. 1. If $\alpha = 0$, then

$$x = p$$

and the line becomes parallel to the Y-axis.

Cor. 2. If $\alpha = 90°$, then

$$y = p$$

and the line becomes parallel to the X-axis.

22. *If* $x \cos \alpha + y \sin \alpha = p$ *be the equation of a given line, then* $x \cos \alpha + y \sin \alpha = p \pm d$ *is the equation of a parallel line.* For the perpendiculars p and $p \pm d$ coincide in direction since they have the same directional cosines; hence the lines to which they are perpendicular are parallel.

Cor. 1. Since

$$p \pm d - p = \pm d$$

it is evident that d is the distance between the lines. If, therefore, (x', y') be a point on the line whose distance from the origin is $p \pm d$, we have

$$x' \cos \alpha + y' \sin \alpha = p \pm d.$$
$$\therefore \pm d = x' \cos \alpha + y' \sin \alpha - p \;\; \cdots \;\; (1)$$

Hence the distance of a point (x', y') from the line $x \cos \alpha + y \sin \alpha = p$ is found by transposing the constant term to the first member, and substituting for x and y the co-ordinates x', y' of the point. Let us, for example, find the distance of the point $(\sqrt{3}, 9)$ from the line $x \cos 30° + y \sin 30° = 5$.

From (1)
$$d = \sqrt{3} \cos 30° + 9 \sin 30° - 5$$
$$= \sqrt{3} \cdot \frac{\sqrt{3}}{2} + 9 \cdot \frac{1}{2} - 5$$
$$\therefore d = 1.$$

From Fig. 13 we have $\cos \alpha = \dfrac{p}{a}$, $\sin \alpha = \dfrac{p}{b} = \dfrac{a}{\sqrt{a^2 + b^2}}$

$\therefore p = \dfrac{ab}{\sqrt{a^2 + b^2}}$

Hence $\pm d = \left(\dfrac{x'}{a} + \dfrac{y'}{b} - 1 \right) \dfrac{ab}{\sqrt{a^2 + b^2}}$.

is the expression for the distance of the point (x', y') from a line whose equation is of the form $\dfrac{x}{a} + \dfrac{y}{b} = 1$.

Let the student show that the expression for d becomes

$$d = \frac{Ax' + By' + C}{\sqrt{A^2 + B^2}}$$

when the equation of the line is given in its general form. See Art. 24, Equation (1).

EXAMPLES.

1. The perpendicular let fall from the origin on a straight line $= 5$ and makes an angle of $30°$ with X-axis; required the equation of the line.

Ans. $\sqrt{3}\,x + y = 10$.

2. The perpendicular from the origin on a straight line makes an angle of $45°$ with the X-axis and its length $= \sqrt{2}$; required the equation of the line.

Ans. $x + y = 2$.

3. What is the distance of the point $(2, 4)$ from the line

$\dfrac{x}{4} + \dfrac{y}{2} = 1$.   *Ans.* $d = \dfrac{6}{\sqrt{5}}$.

Find the distance of the point from the line in each of the following cases:

4. From $(2, 5)$ to $\dfrac{x}{3} - \dfrac{y}{2} = 1$.

5. From $(3, 0)$ to $\dfrac{x}{4} - \dfrac{y}{3} = 1$.

6. From $(0, 1)$ to $2\,y - x = 2$.

7. From (a, c) to $y = sx + b$.

23. *To find the equation of a straight line referred to oblique axes, given the angle between the axes, the angle which the line makes with the X-axis and its Y-intercept.*

NOTE. — Oblique axes are those which intersect at oblique angles.

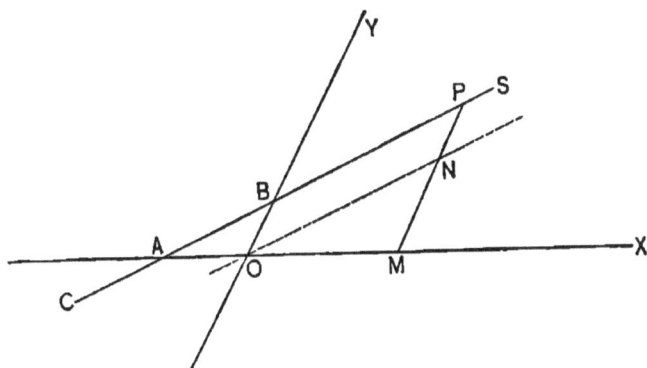

FIG. 14.

Let CS be the line whose equation we wish to determine, it being *any* line in the plane YOX.

Let YOX $= \beta$, SAX $= \alpha$, OB $= b$.

Take any point P on the line and draw

PM ∥ to OY and ON ∥ to SC;

then, PM $= y$, OM $= x$, NOX $= \alpha$, NP $=$ OB $= b$.

From the figure

$$y = MN + NP = MN + b \dots (1)$$

From triangle ONM, we have,

$$\frac{MN}{OM} = \frac{\sin NOM}{\sin MNO};$$

$$\therefore \frac{MN}{x} = \frac{\sin \alpha}{\sin (\beta - \alpha)}.$$

Substituting the value of MN drawn from this equation in (1), we have,

$$y = \frac{\sin \alpha}{\sin (\beta - \alpha)} x + b \dots (2)$$

This equation expresses the relationship between the co-ordinates of at least one point on the line. But as the point selected was *any* point, the above relation holds good for *every* point, and is, therefore, the algebraic expression of the law which governed the motion of the moving point in describing the line. It is therefore the equation of the line.

Cor. 1. If $b = 0$, then

$$y = \frac{\sin \alpha}{\sin (\beta - \alpha)} x \ . \ . \ . \ (3)$$

is the general equation of a line referred to oblique axes passing through the origin.

Cor. 2. If $b = 0$ and $\alpha = 0$, then

$$y = 0 \ . \ . \ . \ (4)$$

the equation of the X-axis.

Cor. 3. If $b = 0$ and $\beta = \alpha$, then

$$x = 0 \ . \ . \ . \ (5)$$

the equation of the Y-axis.

Cor. 4. If $\beta = 90°$; i.e., if the axes are made rectangular, then

$$y = \tan \alpha \, x + b.$$

But $\tan \alpha = s \ \therefore \ y = sx + b.$

This is the slope equation heretofore deduced. See Art. 19 (1).

Cor. 5. If $\beta = 90°$ and $b = 0$, then

$$y = sx. \quad \text{See Art. 19, Cor. 1.}$$

EXAMPLES.

1. Find the equation of the straight line which makes an angle of 30° with the X-axis and cuts the Y-axis two units distant from the origin, the axes making an angle of 60° with each other.

Ans. $y = x + 2.$

2. If the axes had been assumed rectangular in the example above, what would have been the equation?

Ans. $y = \dfrac{x}{\sqrt{3}} + 2.$

3. The co-ordinate axes are inclined to each other at an angle of 30°, and a line passing through the origin is inclined to the X-axis at an angle of 120°, required the equation of the line.

$$\text{Ans.} \quad y = -\frac{x}{2}\sqrt{3} \cdot$$

24. *Every equation of the first degree between two variables is the equation of a straight line.*

Every equation of the first degree between two variables can be placed under the form

$$Ax + By + C = 0 \ldots (1)$$

in which A, B, and C may be either finite or zero.

Suppose A, B, and C are not zero. Solving with respect to y, we have,

$$y = -\frac{A}{B}x - \frac{C}{B} \ldots (2)$$

Comparing equation (2) with (1) Art. 19, we see that it is the equation of a straight line whose Y-intercept $b = -\frac{C}{B}$ and whose slope $s = -\frac{A}{B}$; hence (1), the equation from which it was derived is the equation of a straight line.

If $A = 0$, then $y = -\frac{C}{B}$,

the equation of a line parallel to the X-axis.

If $B = 0$, then $x = -\frac{C}{A}$,

the equation of a line parallel to the Y-axis.

If $C = 0$, then $y = -\frac{A}{B}x$,

the equation of a line passing through the origin.

Hence, for all values of A, B, C equation (1) is the equation of a straight line.

25. *To find the equation of a straight line passing through a given point.*

Let (x', y') be the given point.

Since the line is to be straight, its equation must be

$$y = sx + b \ . \ . \ . \ (1)$$

in which s and b are to be determined.

Now, the equation of a line expresses the relationship which exists between the co-ordinates of *every* point on it; hence its equation must be satisfied when the co-ordinates of *any* point on it are substituted for the general co-ordinates x and y. We have, therefore, the equation of condition.

$$y' = sx' + b \ . \ . \ . \ (2)$$

But a straight line cannot in general be made to pass through a given point (x', y'), cut off a given distance (b) on the Y-axis, and make a given angle (tan. $= s$) with the X-axis. We must therefore eliminate one of these requirements. By subtracting (2) from (1), we have,

$$y - y' = s\,(x - x') \ . \ . \ . \ (3)$$

which is the required equation.

Cor. 1. If $x' = 0$, then

$$y - y' = sx \ . \ . \ . \ (4)$$

is the equation of a line passing through a point on the Y-axis.

Cor. 2. If $y' = 0$, then

$$y = s\,(x - x') \ . \ . \ . \ (5)$$

is the equation of a line passing through a point on the X-axis.

Cor. 3. If $x' = 0$ and $y' = 0$, then

$$y = sx$$

is the equation (heretofore determined), of a line passing through the origin.

EXAMPLES.

1. Write the equation of several lines which pass through the point (2, 3).

2. What is the equation of the line which passes through $(1, -2)$, and makes an angle whose tangent is 2 with the X-axis ?

Ans. $y = 2\,x - 4$.

3. A straight line passes through $(-1, -3)$, and makes an angle of 45° with the X-axis. What is its equation ?

Ans. $y = x - 2$.

4. Required the equations of the two lines which contain the point (a, b), and make angles of 30° and 60°, respectively with X-axis.

Ans. $y - b = \dfrac{x - a}{\sqrt{3}}$; $y - b = \sqrt{3} \cdot (x - a)$.

26. *To find the equation of a straight line passing through two given points.*

Let (x', y'), (x'', y'') be the given points.

Since the line is straight its equation must be

$$y = sx + b \ \ldots \ (1)$$

in which s and b are to be determined.

Since the line is required to pass through the points (x', y'), (x'', y''), we have the equations of condition.

$$y' = sx' + b \ \ldots \ (2)$$
$$y'' = sx'' + b \ \ldots \ (3)$$

As a straight line cannot, in general, be made to fulfil more than two conditions, we must eliminate two of the four conditions expressed in the three equations above.

Subtracting (2) from (1), and then (3) from (2), we have,

$$y - y' = s\,(x - x')$$
$$y' - y'' = s\,(x' - x'')$$

Dividing these, member by member, we have,

$$\frac{y - y'}{y' - y''} = \frac{x - x'}{x' - x''} .$$

Hence $y - y' = \dfrac{y' - y''}{x' - x''} (x - x')$ . . . (4)

is the required equation.

Cor. 1. If $y' = y''$, then

$$y - y' = 0, \text{ or } y = y',$$

which is, as it should be, the equation of a line ‖ to the X-axis.

Cor. 2. If $x' = x''$, then

$$x - x' = 0, \text{ or } x = x',$$

which is the equation of a line ‖ to the Y-axis.

EXAMPLES.

1. Given the two points $(-1, 6)$, $(-2, 8)$; required both the slope and symmetrical equation of the line passing through them.

Ans. $y = -2x + 4, \dfrac{x}{2} + \dfrac{y}{4} = 1.$

2. The vertices of a triangle are $(-2, 1)$, $(-3, -4)$ $(2, 0)$; required the equations of its sides.

Ans. $\begin{cases} y = 5x + 11 \\ 4x - 5y = 8 \\ 4y + x = 2. \end{cases}$

Write the equations of the lines passing through the points :

3. $(-2, 3), (-3, -1)$
Ans. $y = 4x + 11.$

6. $(5, 2), (-2, 4)$
Ans. $7y + 2x = 24.$

4. $(1, 4), (0, 0)$
Ans. $y = 4x.$

7. $(2, 0), (-3, 0)$
Ans. $y = 0.$

5. $(0, 2), (3, -1)$
Ans. $y + x = 2.$

8. $(-1, -3), (-2, 4)$
Ans. $y + 7x + 10 = 0.$

27. *To find the length of a line joining two given points.*

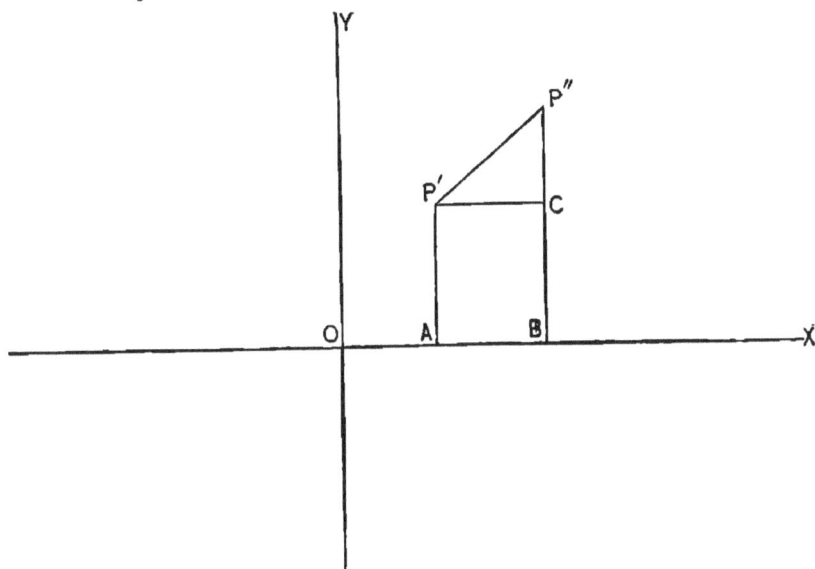

Fig. 15.

Let (x', y'), (x'', y'') be the co-ordinates of the given points P', P''. $L = P'P'' =$ required length.

Draw P''B and P'A $\parallel$ to OY, and P'C $\parallel$ to OX.

We see from the figure that L is the hypothenuse of a right angled triangle whose sides are

$$P'C = AB = OB - OA = x'' - x', \text{ and}$$
$$P''C = P''B - BC = y'' - y'.$$

Hence,

$$P'P'' = L = \sqrt{(x'' - x')^2 + (y'' - y')^2} \cdots (1)$$

Cor. 1. If $x' = 0$ and $y' = 0$, the point P' coincides with the origin, and we have

$$L = \sqrt{x''^2 + y''^2} \cdots (2)$$

for the distance of a point from the origin.

EXAMPLES.

1. Given the points $(2, 0)$, $(-2, 3)$; required the distance between them; also the equation of the line passing through them.

Ans. $L = 5$, $4y + 3x = 6$.

2. The vertices of a triangle are (2, 1) (− 1, 2) (− 3, 0);
what are the lengths of its sides?

Ans. $\sqrt{8}$, $\sqrt{10}$, $\sqrt{26}$.

Give the distances between the following points:

3. (2, 3), (1, 0)   **7.** (− 3, 2), (0, 1)

Ans. $\sqrt{10}$.

4. (4, − 5), (6, − 1)   **8.** (− 2,−1), (2, 0)

Ans. $\sqrt{20}$.

5. (0, 2), (− 1, 0)   **9.** (a, b), (c, d)

Ans. $\sqrt{5}$.

6. (0, 0), (2, 0)   **10.** (− 2, 3), (− a, b).

Ans. 2.

11. What is the expression for the area of a triangle whose
vertices are (x', y'), (x'', y''), (x''', y''') ?

Ans. Area $= \frac{1}{2}\left[x'\left(y''-y'''\right) + x''\left(y'''-y'\right) + x'''\left(y'-y''\right)\right]$.

28. *To find the intersection of two lines given by their
equations.*

Let $y = sx + b$, and
$$y = s'x + b'$$

be the equations of the given lines.

Since each of these equations is satisfied for the co-ordinates
of *every* point on the locus it represents, they must *at the
same time* be satisfied for the co-ordinates of their point of
intersection, as this point is common to both. Hence, for the
co-ordinates of this point the equations are *simultaneous*. So
treating them, we find

$$x = \frac{b - b'}{s' - s}, \text{ and } y = \frac{s'b - sb'}{s' - s}.$$

for the co-ordinates of the required point.

EXAMPLES.

1. Find the intersection of $y = 2x + 1$ and $2y = x - 4$.

Ans. (− 2,−3).

2. The equations of the sides of a triangle are

$$2\,y = 3\,x + 1, \; y + x = 1, \; 2\,y + 4\,x = -3\,;$$

required the co-ordinates of its vertices.

Ans. $\left(\dfrac{1}{5}, \dfrac{4}{5}\right)$, $\left(-\dfrac{4}{7}, -\dfrac{5}{14}\right)$, $\left(-\dfrac{5}{2}, \dfrac{7}{2}\right)$.

3. Write the equation of the line which shall pass through the intersection of $2\,y + 3\,x + 2 = 0$ and $3\,y - x - 8 = 0$, and make an angle with the X-axis whose tangent is 4.

Ans. $y = 4\,x + 10$.

4. What are the equations of the diagonals of the quadrilateral the equations of whose sides are $y - x + 1 = 0$, $y = -x + 2$, $y = 3\,x + 2$, and $y + 2\,x + 2 = 0$?

Ans. $23\,y - 9\,x + 2 = 0, \; 3\,y - 30\,x = 6$.

5. The equation of a chord of the circle whose equation is $x^2 + y^2 = 10$ is $y = x + 2$; required the length of the chord.

Ans. $\text{L} = \sqrt{32}$.

29. If $\qquad \text{A}x + \text{B}y + \text{C} = 0 \;\ldots\; (1)$

and $\qquad \text{A}'x + \text{B}'y + \text{C}' = 0 \;\ldots\; (2)$

be the equations of two straight lines, then

$$\text{A}x + \text{B}y + \text{C} + \text{K}\,(\text{A}'x + \text{B}'y + \text{C}') = 0 \;\ldots\; (3)$$

(K being any constant quantity) is the equation of a straight line which passes through the intersection of the lines represented by (1) and (2). It is the equation of a straight line because it is an equation of the first degree between two variables. See Art. 24. It is also the equation of a straight line which passes through the intersection of (1) and (2), since it is obviously satisfied for the values of x and y which *simultaneously* satisfy (1) and (2).

Let us apply this principle to find the equation of the line which contains the point (2, 3) and which passes through the intersection of $y = 2\,x + 1$ and $2\,y + x = 2$.

From (3) we have $y - 2x - 1 + K(2y + x - 2) = 0$ for the equation of a line which passes through the intersection of the given lines. But by hypotheses the point $(2, 3)$ is on this line; hence $3 - 4 - 1 + K(6 + 2 - 2) = 0$

$$\therefore K = \frac{1}{3}.$$

Substituting this value for K we have,

$$y - 2x - 1 + \frac{1}{3}(2y + x - 2), = 0$$

or, $y - x - 1 = 0$

for the required equation. Let the student verify this result by finding the intersection of the two lines and then finding the equation of the line passing through the two points.

30. *To find the angle between two lines given by their equations.*

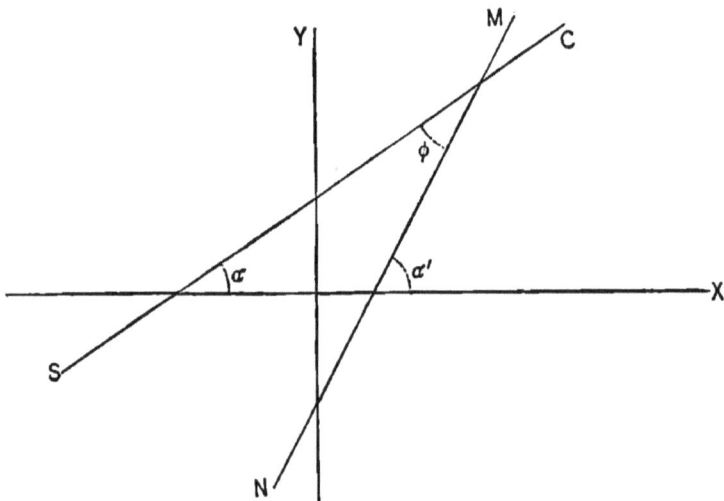

Fig. 16.

Let $y = sx + b$, and

                $y = s'x + b'$

be the equations of SC and MN, respectively; then

        $s = \tan \alpha$ and $s' = \tan \alpha'$.

From the figures

$$\alpha' = \varphi + \alpha$$
$$\therefore \varphi = \alpha' - \alpha.$$

From trigonometry,

$$\tan \varphi = \tan (\alpha' - \alpha) = \frac{\tan \alpha' - \tan \alpha}{1 + \tan \alpha \tan \alpha'}.$$

∴ substituting

$$\tan \varphi = \frac{s' - s}{1 + ss'} \cdots (1).$$

Or,
$$\varphi = \tan^{-1} \frac{s' - s}{1 + ss'} \cdots (2).$$

Cor. 1. If $s = s'$, then

$$\varphi = \tan^{-1} 0 \therefore \varphi = 0.$$

∴ the lines are parallel.

Cor. 2. If $1 + ss' = 0$, then

$$\varphi = \tan^{-1} \infty \therefore \varphi = 90°$$

∴ the lines are perpendicular.

Schol. These results may be obtained geometrically. If the lines are parallel, then, Fig. 16,

$$\alpha = \alpha' \therefore s = s'.$$

If they are perpendicular

$$\alpha' = 90° + \alpha$$

$$\therefore \tan \alpha' = s' = \tan (90° + \alpha) = -\cot \alpha = -\frac{1}{\tan \alpha}.$$

$$\therefore 1 + ss' = 1 + \tan \alpha \tan \alpha' = 1 + \tan \alpha \left(-\frac{1}{\tan \alpha} \right) =$$
$$1 - 1 = 0.$$

EXAMPLES.

1. What is the angle formed by the lines $y - x - 1 = 0$ and $2y + 2x + 1 = 0$?

Ans. $\varphi = 90°.$

2. Required the angle formed by the lines $y + 3x - 2 = 0$ and $2y + 6x + 8 = 0$.

$$Ans. \quad \varphi = 0.$$

3. Required the equation of the line which passes through $(2, -1)$ and is

(*a*) Parallel to $2y - 3x - 5 = 0$.

(*b*) Perpendicular to $2y - 3x - 5 = 0$.

$$Ans. \quad (a) \ 3x - 2y = 8, \quad (b) \ 3y + 2x = 1.$$

4. Given the equations of the sides of a triangle

$$y = 2x + 1, \ y = -x + 2 \text{ and } y = -3; \text{ required.}$$

(*a*) The angles of the triangle.

(*b*) The equations of the perpendiculars from vertices to sides.

(*c*) The lengths of the perpendiculars.

5. What relation exists between the following lines:

$$y = sx + b.$$
$$y = sx - 3.$$
$$y = sx + 6.$$
$$y = sx + m.$$

6. What relation exists between the following:

$$y = sx + b.$$
$$y = -sx + c.$$

7. Find the co-ordinates of the point in which a perpendicular through $(-2, 3)$ intersects $y - 2x + 1 = 0$.

$$Ans. \quad \left(\frac{6}{5}, \frac{7}{5} \right).$$

8. Find the length of the perpendicular let fall from the origin on the line $2y + x = 4$.

$$Ans. \quad L = \frac{1}{5}\sqrt{80}.$$

9. If $Ax + By + C = 0$, $A'x + B'y + C' = 0$, and $A''x + B''y + C'' = 0$ be the equations of three straight lines, and l, m, and n be three constants which render the equation

l (Ax + By + C) + m (A$'x$ + B$'y$ + C$'$) + n (A$''x$ + B$''y$ + C$''$) = 0 an identity, then the three lines meet in a point.

10. Find the equation of the bisector of the angle between the two lines Ax + By + C = 0 and A$'x$ + B$'y$ + C$'$ = 0.

$$\textit{Ans.} \quad \frac{Ax + By + C}{\sqrt{A^2 + B^2}} = \pm \frac{(A'x + B'y + C')}{\sqrt{A'^2 + B'^2}}.$$

GENERAL EXAMPLES.

1. A straight line makes an angle of 45° with the X-axis and cuts off a distance = 2 on the Y-axis; what is its equation when the axes are inclined to each other at an angle of 75° ?

$$\textit{Ans.} \quad y = \sqrt{2}\,x + 2.$$

2. Prove that the lines $y = x + 1$, $y = 2x + 2$ and $y = 3x + 3$ intersect in the point $(-1, 0)$.

3. If (x', y') and (x'', y'') are the co-ordinates of the extremities of a line, show that $\left(\dfrac{x'' + x'}{2}, \dfrac{y'' + y'}{2} \right)$ are the co-ordinates of its middle point.

4. The equations of the sides of a triangle are $y = x + 1$, $x = 4$, $y = -x - 1$; required the equations of the sides of the triangle formed by joining the middle points of the sides of the given triangle.

$$\textit{Ans.} \quad \begin{cases} y = -x + 4 \\ y = x - 4 \\ 2x = 3. \end{cases}$$

5. Prove that the perpendiculars erected at the middle points of the sides of a triangle meet in a common point.

NOTE. — Take the origin at one of the vertices and make the X-axis coincide with one of the sides. Find the equations of the sides; and then find the equations of the perpendiculars at the middle points of the sides. The point of intersection of any two of these perpendiculars ought to satisfy the equation of the third.

6. Prove that the perpendiculars from the vertices of a triangle to the sides opposite meet in a point.

7. Prove that the line joining the middle points of two of the sides of a triangle is parallel to the third side and is equal to one-half of it.

8. The co-ordinates of two of the opposite vertices of a square are $(2, 1)$ and $(4, 3)$; required the co-ordinates of the other two vertices and the equations of the sides.

$\quad$ *Ans.* $(4, 1)$, $(2, 3)$; $y = 1$, $y = 3$, $x = 2$, $x = 4$.

9. Prove that the diagonals of a parallelogram bisect each other.

10. Prove that the diagonals of a rhombus bisect each other at right angles.

11. Prove that the diagonals of a rectangle are equal.

12. Prove that the diagonals of a square are equal and bisect each other at right angles.

13. The distance between the points (x, y) and $(1, 2)$ is $= 4$; give the algebraic expression of the fact.

$\quad$ *Ans.* $(x - 1)^2 + (y - 2)^2 = 4^2$.

14. The points $(1, 2)$, $(2, 3)$ are equi-distant from the point (x, y). Express the fact algebraically.

$$(x - 1)^2 + (y - 2)^2 = (x - 2)^2 + (y - 3)^2; \text{ or, } x + y = 4.$$

15. A circle circumscribes the triangle whose vertices are $(3, 4)$, $(1, -2)$, $(-1, 2)$; required the co-ordinates of its centre.

$\quad$ *Ans.* $(2, 1)$.

16. What is the expression for the distance between the points (x'', y''), (x', y'), the co-ordinate axes being inclined at an angle β?

Ans. $L = \sqrt{(x'' - x')^2 + (y'' - y')^2 + 2(x'' - x')(y'' - y')\cos\beta}$.

17. Given the perpendicular distance (p) of a straight line from the origin and the angle (a) which the perpendicular makes with the X-axis; required the polar equation of the line.

$$Ans. \quad r = \frac{p}{\cos(\theta - a)}.$$

18. Required the length of the perpendicular from the origin on the line $\frac{x}{3} + \frac{y}{4} = 1$. *Ans.* 2.4

19. What is the equation of the line which passes through the point (1, 2), and makes an angle of 45° with the line whose equation is $y + 2x = 1$?

$$Ans. \quad y = -\frac{1}{3}x + \frac{7}{3}.$$

20. One of two lines passes through the points (1, 2), $(-4, -3)$, the other passes through the point $(1, -3)$, and makes an angle of 45° with the first line; required the equations of the lines.

Ans. $y = x + 1$, and $y = -3$, or $x = 1$.

21. If $p = 0$ in the normal equation of a line, through what point does the line pass, and what does its equation become? *Ans.* (0, 0); $y = s\,x$.

22. Required the perpendicular distance of the point ($r\cos\theta$, $r\sin\theta$), from the line $x\cos\theta + y\sin\theta = p$. *Ans.* $r - p$.

23. Given the base of a triangle $= 2\,a$, and the difference of the squares of its sides $= 4\,c^2$. Show that the locus of the vertex is a straight line.

24. What are the equations of the lines which pass through the origin, and divide the line joining the points (0, 1), (1, 0), into three equal parts. *Ans.* $2x = y$, $2y = x$.

25. If (x', y') and (x'', y'') be the co-ordinates of two points, show that the point $\left(\dfrac{mx'' + nx'}{m + n}, \dfrac{my'' + ny'}{m + n}\right)$ divides the line joining them into two parts which bear to each the ratio $m : n$.

CHAPTER IV.

TRANSFORMATION OF CO-ORDINATES.

31. It frequently happens that the discussion of an equation and the deduction of the properties of the locus it represents are greatly simplified by changing the position of the axes to which the locus is referred, thus simplifying the equation, or reducing it to some desired form. The operation by which this is accomplished is termed the TRANSFORMATION OF CO-ORDINATES.

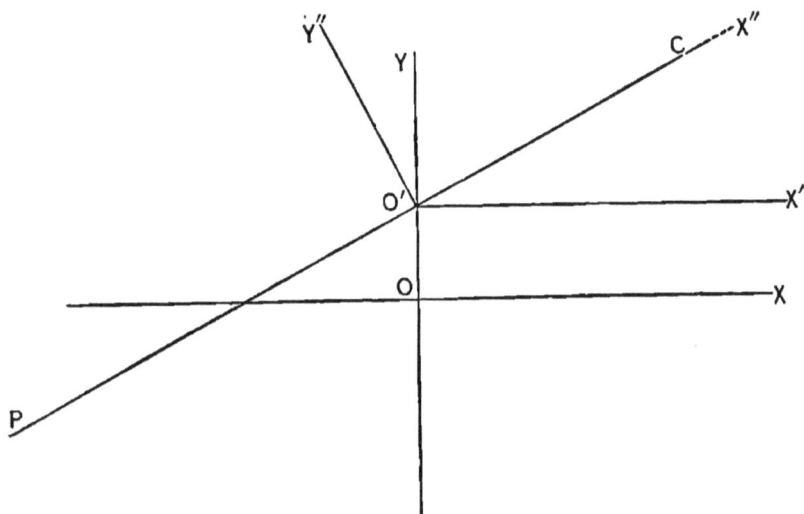

FIG. 17.

The equation of the line PC, Fig. 17, is

$$y = sx + b$$

when referred to the axes Y and X. If we refer it to the axes Y and X' its equation takes the simpler form

$$y = sx'.$$

If we refer it to Y″ and X″, the equation assumes the yet simpler form

$$y'' = 0.$$

Hence, it appears that the position of the axes materially affects the form of the equation of a locus referred to them.

NOTE. — The equation of a locus which is referred to rectangular co-ordinates is called the RECTANGULAR EQUATION of the locus; when referred to polar co-ordinates, the equation is called the POLAR EQUATION of the locus.

32. *To find the equation of transformation from one system of co-ordinates to a parallel system, the origin being changed.*

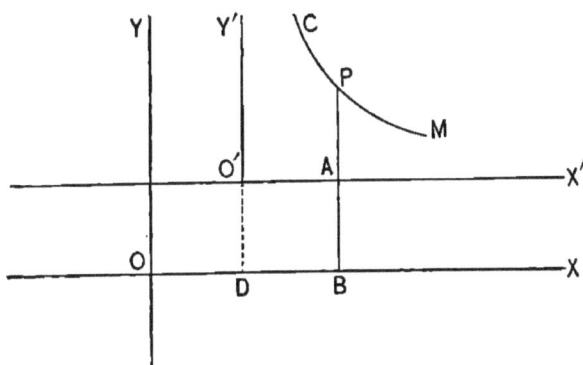

FIG. 18.

Let CM be any plane locus referred to X and Y as axes, and let P be *any* point on that locus. Draw PB ∥ to OY; then from the figure, we have,

(OB, BP) = (x, y) for the co-ordinates of P when referred to X and Y ;

(O′A, AP) = (x', y') for the co-ordinates of P when referred to X′ and Y′;

(OD, DO′) = (a, b) for the co-ordinates of O′, the new origin.

From the figure $OB = OD + DB$; and $BP = BA + AP$;
hence $\qquad x = a + x'$ and $y = b + y'$
are the desired equations.

As these equations express the relations between x, a, x', y, b, and y' for *any* point on the locus they express the relations between the quantities for *every* point. Hence, since the equation of the locus CM expresses the relationship between the co-ordinates of every point on it if we substitute for x and y in that equation their values in terms of x' and y' the resulting or *transformed* equation will express the relationship between the x' and y' co-ordinates for every point on it.

EXAMPLES.

1. What does the equation $y = 3x + 1$ become when the origin is removed to $(2, 3)$?

Ans. $y = 3x + 4$.

2. Construct the locus of the equation $2y - x = 2$. Transfer the origin to $(1, 2)$ and re-construct.

3. The equation of a curve is $y^2 + x^2 + 4y - 4x - 8 = 0$; what does the equation become when the origin is taken at $(2, -2)$?

Ans. $x^2 + y^2 = 16$.

4. What does the equation $y^2 - 2x^2 - 2y + 6x - 3 = 0$ become when the origin is removed to $\left(\dfrac{3}{2}, 1 \right)$?

Ans. $2y^2 - 4x^2 = -1$.

5. The equation of a circle is $x^2 + y^2 = a^2$ when referred to rectangular axes through the centre. What does this equation become when the origin is taken at the left-hand extremity of the horizontal diameter ?

33. *To find the equations of transformation from a rectangular system to an oblique system, the origin being changed.*

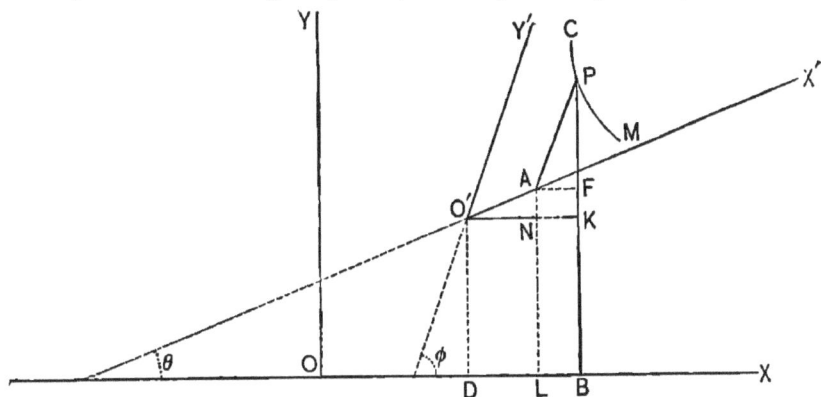

Fig. 19.

Let P be any point on the locus CM.

Let O'Y', O'X' be the new axes, making the angles φ and θ with the X-axis. Draw PA ∥ to the Y'-axis; also the lines

O'D, AL, PB ∥ to the Y-axis, and

AF, O'K ∥ to the X-axis.

From the figure, we have,

$$OB = OD + O'N + AF, \text{ and}$$
$$PB = DO' + AN + PF.$$

But $\qquad OB = x, \ OD = a, \ O'N = x' \cos \theta, \ AF = y' \cos \varphi,$

$$PB = y, \ DO' = b, \ AN = x' \sin \theta, \ PF = y' \sin \varphi;$$

hence, substituting, we have,

$$\left.\begin{array}{l} x = a + x' \cos \theta + y' \cos \varphi \\ y = b + x' \sin \theta + y' \sin \varphi \end{array}\right\} \ \cdots \ (1)$$

for the required equations.

Cor. 1. If $a = 0$, and $b = 0$, O' coincides with O, and we have,

$$\left.\begin{array}{l} x = x' \cos \theta + y' \cos \varphi \\ y = x' \sin \theta + y' \sin \varphi \end{array}\right\} \ \cdots \ (2)$$

for the equations of transformation from a rectangular system to an oblique system, the origin remaining the same.

Cor. 2. If $a = 0$, $b = 0$, and $\varphi = 90° + \theta$, O' coincides with O and the new axes X' and Y' are rectangular. Making these substitutions, and recollecting that

$$\cos \varphi = \cos (90° + \theta) = - \sin \theta, \text{ and}$$
$$\sin \varphi = \sin (90° + \theta) = \cos \theta,$$

we have,

$$\left. \begin{aligned} x &= x' \cos \theta - y' \sin \theta \\ y &= x' \sin \theta + y' \cos \theta \end{aligned} \right\} \quad \cdots \quad (3)$$

for the equations of transformation from one rectangular system to another rectangular system, the origin remaining the same.

Note. — If we find the values of x' and y' in equations (2) in terms of x and y we obtain the equations of transformation from an oblique system to a rectangular system, the origin remaining the same.

EXAMPLES.

1. What does the equation $x^2 + y^2 = 16$ become when the axes are turned through an angle of 45° ?

Ans. The equation is unchanged.

2. The equation of a line is $y = x - 1$; required the equation of the same line when referred to axes making angles of 45° and 135° with the old axis of x.

Ans. $y = - \sqrt{\frac{1}{2}}.$

3. What does the equation of the line in Example 2 become when referred to the old Y-axis and a new X-axis, making an angle of 30° with the old X-axis.

Ans. $2 y = (\sqrt{3} - 1) x - 2.$

34. *To find the equations of transformation from a rectangular system to a polar system, the origin and pole non-coincident.*

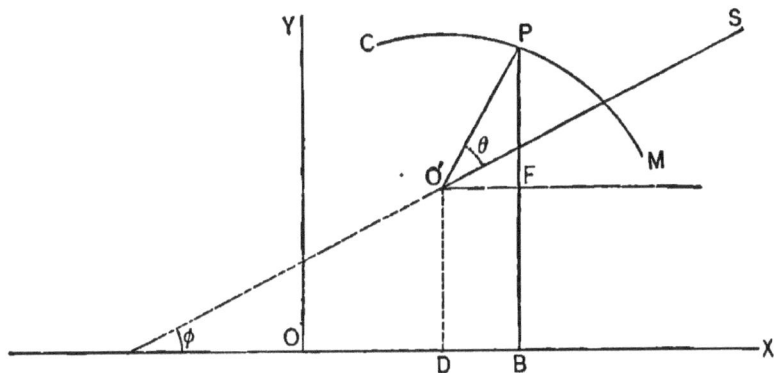

FIG. 20.

Let O' (*a*, *b*) be the pole and O'S the initial line, making an angle φ with the X-axis. Let CM be any locus and P any point on it. From the figure, we have,

$$OB = OD + O'F,$$
$$BP = DO' + FP.$$

But $OB = x$, $OD = a$, $O'F = O'P \cos PO'F = r \cos (\theta + \varphi)$
$BP = y$, $DO' = b$, $FP = O'P \sin PO'F = r \sin (\theta + \varphi)$;

hence, substituting, we have,

$$\left. \begin{array}{l} x = a + r \cos (\theta + \varphi) \\ y = b + r \sin (\theta + \varphi) \end{array} \right\} \quad \cdots \quad (1)$$

for the required equations.

Cor. 1. If the initial line O'S is parallel to the X-axis (it is usual to so take it) $\varphi = 0$, and

$$\left. \begin{array}{l} x = a + r \cos \theta \\ y = b + r \sin \theta \end{array} \right\} \quad \cdots \quad (2)$$

become the equations of transformation.

Cor. 2. If the pole is taken at the origin O, and the initial line made coincident with the X-axis $a = 0$, $b = 0$, and $\varphi = 0$.

Hence, in this case,

$$\left.\begin{array}{l} x = r \cos \theta \\ y = r \sin \theta \end{array}\right\} \quad \dots \quad (3)$$

will be the required equations of transformation.

35. *To find the equations of transformation from a polar system to a rectangular system.*

1°. *When the pole and origin are coincident, and when the initial line coincides with the X-axis.*

From equations (3), Art. 34, we have, by squaring and adding $r^2 = x^2 + y^2$; and,

by division $\tan \theta = \dfrac{y}{x}$.

for the required equations. We have, also, from the same equations,

$$\cos \theta = \frac{x}{r} = \frac{x}{\sqrt{x^2 + y^2}} \, ; \; \sin \theta = \frac{y}{r} = \frac{y}{\sqrt{x^2 + y^2}} \, .$$

2°. *When the pole and origin are non-coincident, and when the initial line is parallel to the X-axis.*

From equations (2) of the same article, we have, by a similar process,

$$r^2 = (x - a)^2 + (y - b)^2$$

$$\tan. \theta = \frac{y - b}{x - a} \, ; \; \text{also}$$

$$\cos \theta = \frac{x - a}{r} = \frac{x - a}{\sqrt{(x - a)^2 + (y - b)^2}} \, ;$$

$$\sin \theta = \frac{y - b}{r} = \frac{y - b}{\sqrt{(x - a)^2 + (y - b)^2}} \, .$$

for the required equations.

EXAMPLES.

1. The rectangular equation of the circle is $x^2 + y^2 = a^2$; what is its polar equation when the origin and pole are coincident and the initial line coincides with the X-axis ?

Ans. $r = a.$

2. The equation of a curve is $(x^2 + y^2)^2 = a^2 (x^2 - y^2)$; required its polar equation, the pole and initial being taken as in the previous example.

Ans. $r^2 = a^2 \cos 2\,\theta.$

Deduce the rectangular equation of the following curves, assuming the origin at the pole and the initial line coincident with the X-axis.

3. $r = a \tan^2 \theta \sec \theta$

Ans. $x^{\frac{3}{2}} = a^{\frac{1}{2}} y.$

4. $r^2 = a^2 \tan \theta \sec^2 \theta$

Ans. $x^3 = a^2 y.$

5. $r^2 = a^2 \sin 2\,\theta$

Ans. $(x^2 + y^2)^2 = 2\,a^2 xy.$

6. $r = a (\cos \theta - \sin \theta)$

Ans. $x^2 + y^2 = a (x - y)$

GENERAL EXAMPLES.

Construct each of the following straight lines, transfer the origin to the point indicated, the new axes being parallel to the old, and reconstruct :

1. $y = 3\,x + 1$ to $(1, 2)$.

2. $2\,y - x - 2 = 0$ to $(-1, 2)$.

3. $\frac{1}{2} y + x - 4 = 0$ to $(-2, -1)$.

4. $y + x + 1 = 0$ to $(0, 2)$.

5. $y = sx + b$ to (c, d).

6. $y + 2\,x = 0$ to $(2, -2)$.

7. $y = mx$ to (l, n).

8. $y - 4\,x + c = 0$ to (d, o).

What do the equations of the following curves become when referred to a parallel (rectangular) system of co-ordinates passing through the indicated points ?

9. $3\,x^2 + 2\,y^2 = 6$, $(\sqrt{2}, 0)$.

10. $y^2 = 4\,x$ $(1, 0)$.

11. $9\,y^2 - 4\,x^2 = -36$ $(3, 0)$.

12. $y^2 = 2\,px \left(-\frac{p}{2}, 0 \right).$

13. What does the equation $x^2 + y^2 = 4$ become when the X-axis is turned to the left through an angle of $30°$ and the Y-axis is turned to the right through the same angle ?

14. What does the equation $x^2 - y^2 = a^2$ become when the axes are turned through an angle of $-45°$?

15. What is the polar equation of the curve $y^2 = 2\,px$, the pole and origin being coincident, and the initial line coinciding with the X-axis ?

16. The polar equation of a curve is $r = a\,(1 + 2\cos\theta)$; required its rectangular equation, the origin and pole being coincident and the X-axis coinciding with the initial line.

$$Ans. \quad (x^2 + y^2 - 2\,ax)^2 = a^2\,(x^2 + y^2).$$

Required the rectangular equation of the following curves, the pole, origin, initial line, and X-axis being related as in Example 16

17. $r^2 = \dfrac{a^2}{\cos 2\theta}$.   **20.** $r = a \sec^2 \dfrac{\theta}{2}$.

$Ans.$ $x^2 - y^2 = a^2$.

18. $r = a \sin\theta$.   **21.** $r = a \sin 2\theta$.

19. $r = a\,\theta$.   **22.** $r^2 - 2\,r\,(\cos\theta + \sqrt{3}\sin\theta) = 5$.

Find the polar equations of the loci whose rectangular equations are:

23. $x^3 = y^2\,(2\,a - x)$.   **25.** $a^4 y^2 = a^2 x^4 - x^6$.

24. $4a^2 x = y^2\,(2\,a - x)$. **26.** $x^{\frac{2}{3}} + y^{\frac{2}{3}} = a^{\frac{2}{3}}$.

CHAPTER V.

THE CIRCLE.

36. The circle is a curve generated by a point moving in the same plane so as to remain at the same distance from a fixed point. It will be observed that the circle as here defined is the same as the circumference as defined in plane geometry.

37. *Given the centre of a circle and its radius to deduce its equation.*

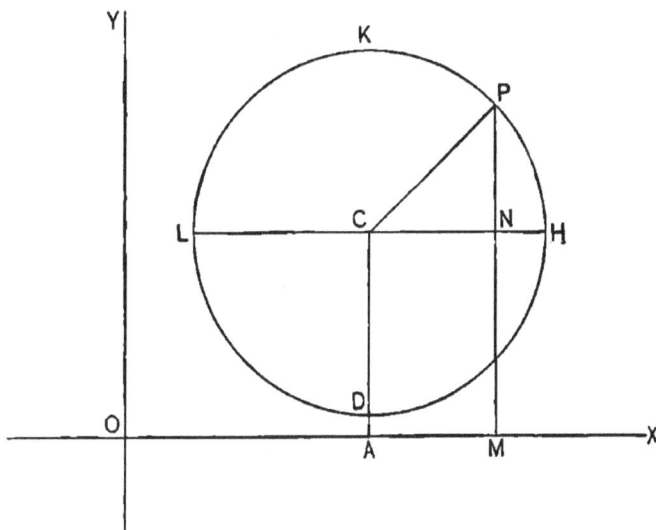

FIG. 21.

Let C (x', y') be the centre of the circle, and let P be any point on the curve. Draw CA and PM ∥ to OY and CN ∥ to OX; then

(OA, AC) = (x', y') are the co-ordinates of the centre C.

(OM, MP) = (x, y) are the co-ordinates of the point P.

Let CP $= a$.   From the figure, we have,

$$CN^2 + NP^2 = CP^2; \ldots (1)$$

But $CN^2 = (OM - OA)^2 = (x - x')^2,$
$NP^2 = (MP - AC)^2 = (y - y')^2,$ and
$CP^2 = a^2$

Substituting these values in (1), we have,

$$(x - x')^2 + (y - y')^2 = a^2 \ldots (2)$$

for the required equation.   For equation (2) expresses the relation existing between the co-ordinates of *any* point (P) on the circle; hence it expresses the relation between the co-ordinates of *every* point.   It is, therefore, the equation of the circle.

If in (2) we make $x' = 0$ and $y' = 0$, we have,

$$x^2 + y^2 = a^2 \ldots (3)$$

or, symmetrically, $\dfrac{x^2}{a^2} + \dfrac{y^2}{a^2} = 1 \ldots (4)$

for the equation of the circle when referred to rectangular axes passing through the centre.

Let the student discuss and construct equation (3).   See Art. 11.

Cor. 1.   If we transpose x^2 in (3) to the second member and factor, we have,

$$y^2 = (a + x) \ (a - x);$$

i.e., *in the circle the ordinate is a mean proportional between the segments into which it divides the diameter.*

Cor. 2.   If we take *L*, Fig. 21, as the origin of co-ordinates, and the diameter *L*H as the X-axis, we have,

$$LC = x' = a \text{ and } y' = 0.$$

These values of x' and y' in (2) give

$$(x - a)^2 + y^2 = a^2,$$

or, after reduction, $x^2 + y^2 - 2\,ax = 0 \ldots (5)$

for the equation of the circle when referred to rectangular

axes taken at the left hand extremity of the horizontal diameter.

38. *Every equation of the second degree between two varia-bles, in which the coefficients of the second powers of the variables are equal and the term in xy is missing, is the equa-tion of a circle.*

The most general equation of the second degree in which these conditions obtain is

$$ax^2 + ay^2 + cx + dy + f = 0. \ \cdots \ (1)$$

Dividing through by a and re-arranging, we have,

$$x^2 + \frac{c}{a}x + y^2 + \frac{d}{a}y = -\frac{f}{a}.$$

If to both members we now add

$$\frac{c^2}{4\,a^2} + \frac{d^2}{4\,a^2},$$

the equation may be put under the form

$$\left(x + \frac{c}{2\,a}\right)^2 + \left(y + \frac{d}{2\,a}\right)^2 = \frac{c^2 + d^2 - 4\,af}{4\,a^2}.$$

Comparing this with (2) of the preceding article, we see that it is the equation of a circle in which

$$\left(-\frac{c}{2\,a}, \ -\frac{d}{2\,a}\right)$$

are the co-ordinates of the centre and

$$\frac{\sqrt{c^2 + d^2 - 4\,af}}{2\,a} \text{ is the radius.}$$

Cor. 1. If $ax^2 + ay^2 + cx + dy + m = 0$ be the equation of another circle, it must be *concentric* with the circle repre-sented by (1); for the co-ordinates of the centre are the same. Hence, when the equations of circles have the variables in their terms affected with equal coefficients, each to each, the circles are concentric. Thus

$$2 x^2 + 2 y^2 + 3 x + 4 y + 9 = 0$$
$$2 x^2 + 2 y^2 + 3 x + 4 y + 25 = 0$$

are the equations of concentric circles.

EXAMPLES.

What is the equation of the circle when the origin is taken.

1. At D, Fig. 21 ?   *Ans.* $x^2 + y^2 - 2 a y = 0.$

2. At K, Fig. 21 ?   *Ans.* $x^2 + y^2 + 2 a y = 0.$

3. At H, Fig. 21 ?   *Ans.* $x^2 + y^2 + 2 a x = 0.$

What are the co-ordinates of the centres, and the values of the radii of the following circles ?

4. $4 x^2 + 4 y^2 - 8 x - 8 y + 2 = 0.$
         Ans. $(1, 1), a = \sqrt{\tfrac{3}{2}}.$

5. $x^2 + y^2 + 4 x - 6 y - 3 = 0.$
         Ans. $(- 2, 3), a = 4.$

6. $2 x^2 + 2 y^2 - 8 x = 0.$
         Ans. $(2, 0), a = 2.$

7. $x^2 + y^2 - 6 x = 0.$
         Ans. $(3, 0), a = 3.$

8. $x^2 + y^2 - 4 x + 8 y - 5 = 0.$
         Ans. $(2, - 4) a = 5.$

9. $x^2 + y^2 - m x + n y + c = 0.$

10. $x^2 + y^2 = m.$

11. $x^2 - 4 x = - y^2 - m y.$

12. $x^2 + y^2 = c^2 + d^2.$

13. $x^2 + c x + y^2 = f.$

Write the equations of the circles whose radii and whose centres are

14. $a = 3$, $(0, 1)$.
Ans. $x^2 + y^2 - 2y = 8$.

15. $a = 2$, $(1, -2)$.
Ans. $x^2 + y^2 - 2x + 4y + 1 = 0$.

16. $a = 5$, $(-2, -2)$.
Ans. $x^2 + y^2 + 4y + 4x = 17$.

17. $a = 4$, $(0, 0)$.
Ans. $x^2 + y^2 = 16$.

18. $a = m$, (b, c).

19. $a = b$, $(c, -d)$.

20. $a = 5$, (l, k).

21. $a = k$, $(2, b)$.

22. The radius of a circle is 5; what is its equation if it is concentric with $x^2 + y^2 - 4x = 2$?
Ans. $x^2 + y^2 - 4x = 21$.

23. Write the equations of two concentric circles which have for their common centre the point $(2, -1)$.

24. Find the equation of a circle passing through three given points.

39. *To deduce the polar equation of the circle.*

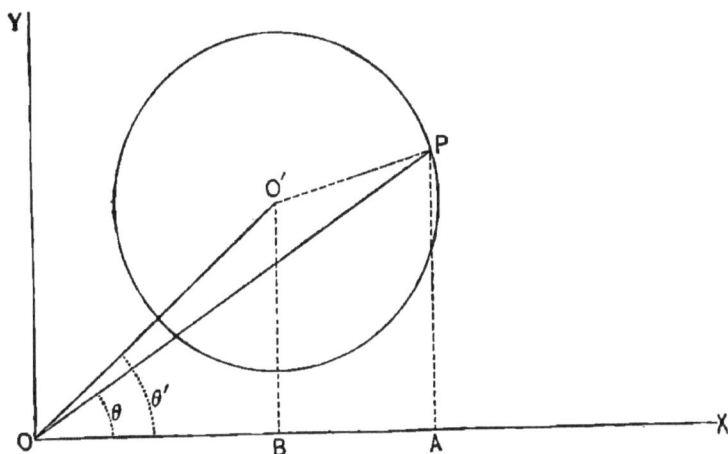

FIG. 22.

The equation of the circle when referred to OY, OX is
$$(x - x')^2 + (y - y')^2 = a^2.$$

To deduce the polar equation let P be any point of the curve, then

$$(OA, AP) = (x, y)$$
$$(OB, BO') = (x', y')$$
$$(OP, POA) = (r, \theta)$$
$$(OO', O'OB) = (r', \theta')$$

From the figure, $OA = x = r \cos \theta$, $AP = y = r \sin \theta$,

$$OB = x' = r' \cos \theta', \quad BO' = y' = r' \sin \theta' ;$$

hence, substituting, we have,

$$(r \cos \theta - r' \cos \theta')^2 + (r \sin \theta - r' \sin \theta')^2 = a^2.$$

Squaring and collecting, we have,

$$r^2(\cos^2 \theta + \sin^2 \theta) + r'^2(\cos^2 \theta' + \sin^2 \theta') - 2 rr'(\cos \theta \cos \theta' + \sin \theta \sin \theta') = a^2$$

i.e., $\qquad r^2 + r'^2 - 2 rr' \cos (\theta - \theta') = a^2 \ldots (1)$

is the polar equation of the circle.

This equation might have been obtained directly from the triangle OO'P.

Cor. 1. If $\theta' = 0$, the initial line OX passes through the centre and the equation becomes

$$r^2 + r'^2 - 2 rr' \cos \theta = a^2.$$

Cor. 2. If $\theta' = 0$, and $r' = a$, the pole lies on the circumference and the equation becomes

$$r = 2 a \cos \theta.$$

Cor. 3. If $\theta' = 0$, and $r' = 0$, the pole is at the centre and the equation becomes

$$r = a.$$

40. *To show that the supplemental chords of the circle are perpendicular to each other.*

The supplemental chords of a circle are those chords which pass through the extremities of any diameter and intersect each other on the circumference.

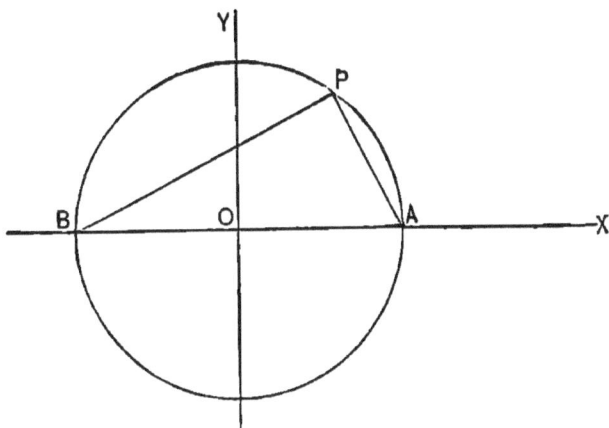

FIG. 23.

Let PB, PA be a pair of supplemental chords. We wish to prove that they are at right angles to each other.
The equation of a line through B (− *a*, *o*) is

$$y = s\,(x + a).$$

For a line through A (*a*, *o*), we have

$$y = s'\,(x - a).$$

Multiplying these, member by member, we have

$$y^2 = ss'\,(x^2 - a^2) \ldots (a)$$

for an equation which expresses the relation between the co-ordinates of the point of intersection of the lines.

Since the lines must not only intersect, but intersect *on* the circle whose equation is

$$y^2 = a^2 - x^2,$$

this equation must subsist at the same time with equation (*a*) above; hence, dividing, we have

$$1 = -\,ss',$$

or, $1 + ss' = 0 \ldots (1)$

Hence the supplemental chords of a circle are perpendicular to each other.

Let the student discuss the proposition for a pair of chords passing through the extremities of the vertical diameter.

41. *To deduce the equation of the tangent to the circle.*

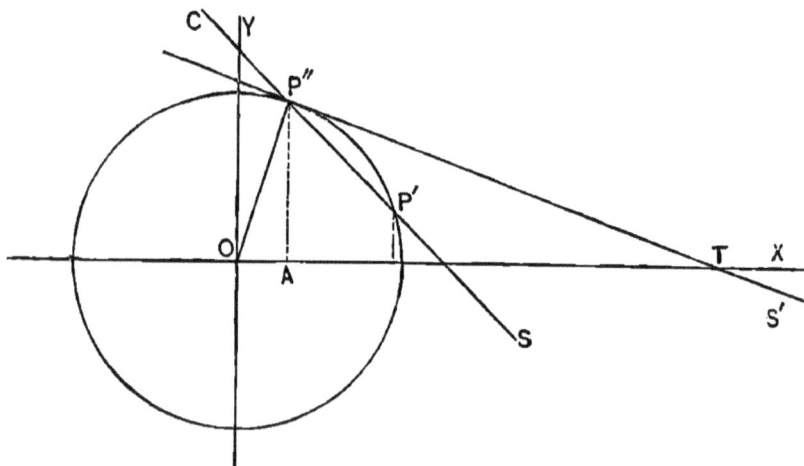

FIG. 24.

Let CS be any line cutting the circle in the points P' (x', y'), P'' (x'', y''). Its equation is

$$y - y' = \frac{y' - y''}{x' - x''} (x - x'). \quad \text{(Art. 26, (4))}.$$

Since the points (x', y'), (x'', y'') are on the circle, we have the equations of condition

$$x'^2 + y'^2 = a^2 \quad \ldots \quad (1)$$
$$x''^2 + y''^2 = a^2 \quad \ldots \quad (2)$$

These three equations must subsist at the same time; hence, subtracting (2) from (1) and factoring, we have,

$$(x' + x'')(x' - x'') + (y' + y'')(y' - y'') = 0;$$
$$\therefore \frac{y' - y''}{x' - x''} = - \frac{x' + x''}{y' + y''}.$$

Substituting in the equation of the secant line it becomes

$$y - y' = - \frac{x' + x''}{y' + y''} (x - x') \quad \ldots \quad (3)$$

If we now revolve the secant line upward about P'' the point P' will approach P'' and will finally coincide with it when the secant CS becomes tangent to the curve. But when

P' coincides with P'', $x' = x''$ and $y' = y''$; hence, substituting in (3) we have,

$$y - y'' = - \frac{x''}{y''} (x - x''), \; \ldots \; (4)$$

or, after reduction,

$$xx'' + yy'' = a^2; \; \ldots \; (5)$$

or, symmetrically,

$$\frac{xx''}{a^2} + \frac{yy''}{a^2} = 1 \; \ldots \; (6)$$

for *the equation of the tangent.*

SCHOL. The SUB-TANGENT for a given point of a curve is the distance from the foot of the ordinate of the point of tangency to the point in which the tangent intersects the X-axis; thus, in Fig. 24, AT is the sub-tangent for the point P''. To find its value make $y = 0$ in the equation of the tangent (5) and we have.

$$OT = x = \frac{a^2}{x''}.$$

But
$$AT = OT - OA = \frac{a^2}{x''} - x''$$

$$\therefore \; \text{sub-tangent} = \frac{a^2 - x''^2}{x''} = \frac{y''^2}{x''}.$$

42. *To deduce the equation of the normal to the circle.*

The normal to a curve at a given point is a line perpendicular to the tangent drawn at that point.

The equation of any line through the point P'' (x'', y'') Fig. 24, is
$$y - y'' = s (x - x'') \; \ldots \; (1)$$

In order that this line shall be perpendicular to the tangent P''T, we must have

$$1 + ss' = 0.$$

But Art. 41, (4) $s' = - \frac{x''}{y''}$; hence, we must have $s = \frac{y''}{x''}$.

Therefore, substituting in (1), we have,

$$y - y'' = \frac{y''}{x''}(x - x'') \; \dots \; (2);$$

or, after reduction,

$$yx'' - xy'' = 0 \; \dots \; (3)$$

for *the equation of the normal.*

We see from the form of this equation that the normal to the circle passes through the centre.

SCHOL. The SUB-NORMAL for a given point on a curve is the distance from the foot of the ordinate of the point to the point in which the normal intersects the X-axis. In the circle, we see from Fig. 24 that the

$$Sub\text{-}normal = x''.$$

43. By methods precisely analogous to those developed in the last two articles, we may prove the equation of the tangent to

$$(x - x')^2 + (y - y')^2 = a^2$$

to be

$$(x - x')(x'' - x') + (y - y')(y'' - y') = a^2 \; \dots \; (1)$$

and that of the normal to be

$$(y - y'')(x'' - x') - (x - x'')(y'' - y') = 0 \; \dots \; (2)$$

Let the student deduce these equations.

EXAMPLES.

1. What is the polar equation of the circle $ax^2 + ay^2 + cx + dy + f = 0$, the origin being taken as the pole and the X-axis as the initial line ?

$$Ans. \quad r^2 + \left(\frac{c}{a} \cos \theta + \frac{d}{a} \sin \theta \right) r + \frac{f}{a} = 0.$$

2. What is the equation of the tangent to the circle $x^2 + y^2 = 25$ at the point $(3, 4)$? The value of the subtangent ?   *Ans.* $3x + 4y = 25$; $\frac{16}{3}$.

3. What is the equation of the normal to the circle $x^2 + y^2 = 37$ at the point $(1, 6)$? What is the value of the sub-normal ?   *Ans.* $y = 6x$; 1.

4. What are the equations of the tangent and normal to the circle $x^2 + y^2 = 20$ at the point whose abscissa is 2 and ordinate negative? Give also the values of the sub-tangent and sub-normal for this point.

$$Ans. \quad 2\,x - 4\,y = 20; \quad 2\,y + 4\,x = 0;$$
$$Sub\text{-}tangent = 8; \quad sub\text{-}normal = 2.$$

Give the equations of the tangents and normals, and the values of the sub-tangents and sub-normals, to the following circles:

5. $x^2 + y^2 = 12$, at $(2, +\sqrt{8})$.

6. $x^2 + y^2 = 25$, at $(3, -4)$.

7. $x^2 + y^2 = 20$, at $(2,$ ordinate $+)$.

8. $x^2 + y^2 = 32$, at (abscissa $+, -4$).

9. $x^2 + y^2 = a^2$, at (b, c).

10. $x^2 + y^2 = m$, at $(1,$ ordinate $+)$.

11. $x^2 + y^2 = k$, at $(2,$ ordinate $-)$.

12. $x^2 + y^2 = 18$, at $(m,$ ordinate $+)$.

13. Given the circle $x^2 + y^2 = 45$ and the line $2\,y + x = 2$; required the equations of the tangents to the circle which are parallel to the line.

$$Ans. \quad \begin{cases} 3\,x + 6\,y = 45. \\ 3\,x + 6\,y = -45. \end{cases}$$

14. What are the equations of the tangents to the circle $x^2 + y^2 = 45$ which are perpendicular to the line $2\,y + x = 2$?

$$Ans. \quad \begin{cases} 3\,y - 6\,x = 45. \\ 6\,x - 3\,y = 45. \end{cases}$$

16. The point $(3, 6)$ lies outside of the circle $x^2 + y^2 = 9$; required the equations of the tangents to the circle which pass through this point.

$$Ans. \quad \begin{cases} x = 3. \\ 4\,y - 3\,x = 15. \end{cases}$$

17. What is the equation of the tangent to the circle $(x - 2)^2 + (y - 3)^2 = 5$ at the point (4, 4) ?

Ans. $2x + y = 12$.

18. The equation of one of two supplementary chords of the circle $x^2 + y^2 = 9$ is $y = \frac{2}{3}x + 2$, what is the equation of the other ?

Ans. $2y + 3x = 9$.

19. Find the equations of the lines which touch the circle $(x - a)^2 + (y - b)^2 = r^2$ and which are parallel to $y = sx + c$.

20. The equation of a circle is $x^2 + y^2 - 4x + 4y = 9$; required the equation of the normal at the point whose abscissa = 3, and whose ordinate is positive.

Ans. $4x - y = 10$.

44. *To find the length of that portion of the tangent lying between any point on it and the point of tangency.*

Let (x_1, y_1) be the point on the tangent. The distance of this point from the centre of the circle whose equation is

$$(x - x')^2 + (y - y')^2 = a^2 \text{ is evidently}$$
$$\sqrt{(x_1 - x')^2 + (y_1 - y')^2}. \text{ See Art. 27, (1).}$$

But this distance is the hypothenuse of a right angled triangle whose sides are the radius a and the required distance d along the tangent; hence

$$d^2 = (x_1 - x')^2 + (y_1 - y')^2 - a^2 \ldots (1)$$

Cor. 1. If $x' = 0$ and $y' = 0$, then (1) becomes

$$d^2 = x_1^2 + y_1^2 - a^2 \ldots (2)$$

as it ought.

45. *To deduce the equation of the radical axis of two given circles.*

The RADICAL AXIS OF TWO CIRCLES *is the locus of a point from which tangents drawn to the two circles are equal.*

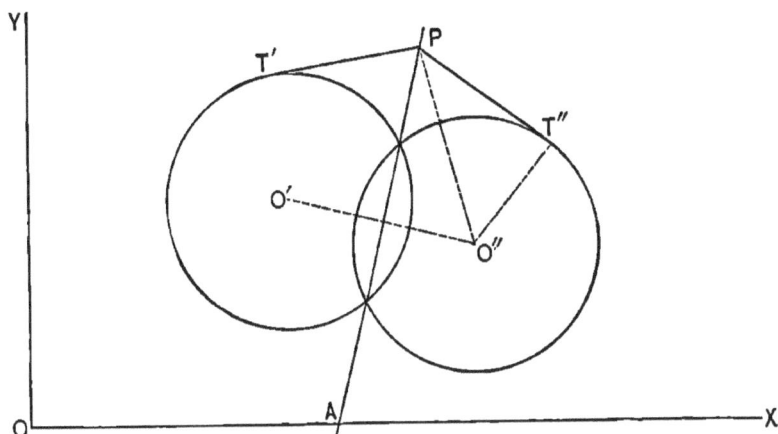

FIG. 25.

Let
$$(x - x')^2 + (y - y')^2 = a^2,$$
$$(x - x'')^2 + (y - y'') = b^2 \text{ be the given circles.}$$

Let P (x_1, y_1) be *any* point on the radical axis; then from the preceding article, we have,

$$d^2 = (x_1 - x')^2 + (y_1 - y')^2 - a^2$$
$$d'^2 = (x_1 - x'')^2 + (y_1 - y'')^2 - b^2$$

∴ by definition $(x_1 - x')^2 + (y_1 - y')^2 - a^2 = (x_1 - x'')^2$
$+ (y_1 - y'')^2 - b^2$; hence, reducing, we have,

$$2(x'' - x') x_1 + 2(y'' - y') y_1 = x''^2 - x'^2 + y''^2 - y'^2$$
$$+ a^2 - b^2.$$

Calling, for brevity, the second member m, we see that (x_1, y_1) will satisfy the equation.

$$2(x'' - x') x + 2(y'' - y') y = m \ \ldots (1)$$

But (x_1, y_1) is *any* point on the radical axis; hence *every* point on that axis will satisfy (1). It is, therefore, the required equation.

COR. 1. If $c = 0$ and $c' = 0$ be the equation of two circles, then, $c - c' = 0$

is the equation of their radical axis.

Cor. 2. From the method of deducing (1) it is easily seen that if the two circles intersect, the co-ordinates of their points of intersection must satisfy (1); hence *the radical axis of two intersecting circles is the line joining their points of intersection,* PA, Fig. 25.

Let the student prove that the radical axis of any two circles is perpendicular to the line joining their centres.

46. *To show that the radical axes of three given circles intersect in a common point.*

Let $c = 0, c' = 0,$ and $c'' = 0$

be the equations of the three circles.

Taking the circles two and two we have for the equations of their radical axes

$$c - c' = 0 \ . \ . \ . \ (1)$$
$$c - c'' = 0 \ . \ . \ . \ (2)$$
$$c' - c'' = 0 \ . \ . \ . \ (3)$$

It is evident that the values of x and y which *simultaneously* satisfy (1) and (2) will also satisfy (3); hence the proposition.

The intersection of the radical axes of three given circles is called THE RADICAL CENTRE *of the circles.*

EXAMPLES.

Find the lengths of the tangents drawn to the following circles :

1. $(x - 2)^2 + (y - 3)^2 = 16$ from $(7, 2)$.

Ans. $d = \sqrt{10}$.

2. $x^2 + (y + 2)^2 = 10$ from $(3, 0)$.

Ans. $d = \sqrt{3}$.

3. $(x - a)^2 + y^2 = 12$ from (b, c).

4. $x^2 + y^2 - 2x + 4y = 2$ from $(3, 1)$.

5. $x^2 + y^2 = 25$ from $(6, 3)$.

Ans. $d = \sqrt{20}$.

6. $x^2 + y^2 - 2x = 10$ from $(5, 2)$.

Ans. $d = 3.$

7. $(x - a)^2 + (y - b)^2 = c$ from (d, f).

8. $x^2 + y^2 - 4y = 10$ from $(0, 0)$.

Give the equations of the radical axis of each of the following pairs of circles :

9. $\begin{cases} (x - 2)^2 + (y - 3)^2 - 10 = 0. \\ (x + 3)^2 + (y + 2)^2 - 6 = 0. \end{cases}$

Ans. $5x + 5y + 2 = 0.$

10. $\begin{cases} x^2 + y^2 - 4y = 0. \\ (x - 3)^2 + y^2 - 9 = 0. \end{cases}$

Ans. $3x = 2y.$

11. $\begin{cases} (x + 3)^2 + y^2 - 2y - 8 = 0. \\ x^2 + y^2 - 2y = 0. \end{cases}$

Ans. $x = -\frac{1}{6}.$

12. $\begin{cases} (x + a)^2 + y^2 - c^2 = 0. \\ x^2 + (y - 3)^2 - 16 = 0. \end{cases}$

13. $\begin{cases} x^2 + y^2 = 16. \\ (x - 1)^2 + y^2 = a^2. \end{cases}$

14. $\begin{cases} x^2 + (y - a)^2 = c^2. \\ (x - 2)^2 + y^2 = d^2. \end{cases}$

Find the co-ordinates of the radical centres of each of the following systems of circles :

15. $\begin{cases} (x - 3)^2 + y^2 = 16. \\ x^2 + y^2 = 9. \\ x^2 + (y - 2)^2 = 25. \end{cases}$

Ans. $(\frac{1}{3}, -3).$

16. $\begin{cases} x^2 + y^2 - 4x + 6y - 3 = 0. \\ x^2 + y^2 - 4x = 12. \\ x^2 + y^2 + 6y = 7. \end{cases}$

Ans. $(1, -\frac{3}{2}).$

17. $\begin{cases} x^2 + y^2 = a. \\ (x - 1)^2 + y^2 = 9. \\ x^2 + y^2 - 2x + 4y = 10. \end{cases}$

18. $\begin{cases} x^2 + y^2 - kx = c. \\ x^2 + y^2 = m. \\ x^2 + y^2 - ay = d. \end{cases}$

47. *To find the condition that a straight line $y = sx + b$ must fulfil in order that it may touch the circle $x^2 + y^2 = a^2$.*

In order that the line may touch the circle the perpendicular let fall from the centre on the line must be equal to the radius of the circle.

From Art. 21, Fig. 13, we have

$$p = b \cos \gamma = \frac{b}{\sec \gamma} = \frac{b}{\sqrt{1 + \tan.^2 \gamma}};$$

$$\therefore p = a = \frac{b}{\sqrt{1 + s^2}};$$

hence, $a^2 (1 + s^2) = b^2 \ldots (1)$

is the required condition.

Cor. 1. If we substitute the value of b drawn from (1) in the equation $y = sx + b$, we have

$$y = sx \pm a \sqrt{1 + s^2} \ldots (2)$$

for the *equation of the tangent in terms of its slope.*

48. *Two tangents are drawn from a point without the circle; required the equation of the chord joining the points of tangency.*

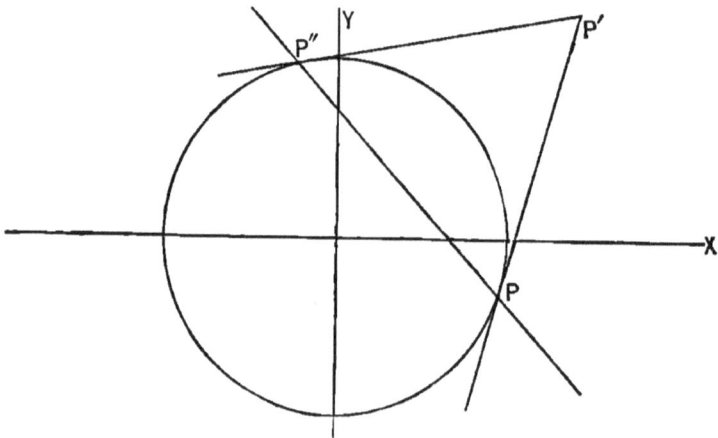

Fig. 26.

Let P' (x', y') be the given point, and let P'P'', P'P, be the tangents through it to the circle.

It is required to deduce the equation of PP″.

The equation of a tangent through P″ (x'', y'') is

$$\frac{xx''}{a^2} + \frac{yy''}{a^2} = 1.$$

Since P′ (x', y') is on this line, its co-ordinates must satisfy the equation; hence

$$\frac{x'x''}{a^2} + \frac{y'y''}{a^2} = 1.$$

The point (x'', y''), therefore, satisfies the equation

$$\frac{x'x}{a^2} + \frac{y'y}{a^2} = 1 ; \ \ldots \ (1)$$

∴ it is a point on the locus represented by (1). A similar course of reasoning will show that P is also a point of this locus. But (1) is the equation of a straight line; hence, since it is satisfied for the co-ordinates of both P″ and P, it is the equation of the straight line joining them. It is, therefore, the required equation.

49. *A chord of a given circle is revolved about one of its points; required the equation of the locus generated by the point of intersection of a pair of tangents drawn to the circle at the points in which the chord cuts the circle.*

Let P′ (x', y'), Fig. 27, be the point about which the chord P′AB revolves. It is required to find the equation of the locus generated by P_1 (x_1, y_1), the intersection of the tangents AP_1, BP_1, as the line P′AB revolves about P′.

From the preceding article the equation of the chord AB is

$$\frac{x_1 x}{a^2} + \frac{y_1 y}{a^2} = 1.$$

Since P′ (x', y') is on this line, we have

$$\frac{x_1 x'}{a^2} + \frac{y_1 y'}{a^2} = 1 ;$$

hence $\quad \dfrac{x'x}{a^2} + \dfrac{y'y}{a^2} = 1 \ \ldots \ (1)$

is satisfied for the co-ordinates of P_1 (x_1, y_1); hence P_1 lies on the locus represented by (1). But P_1 is the intersection of *any* pair of tangents drawn to the circle at the points in

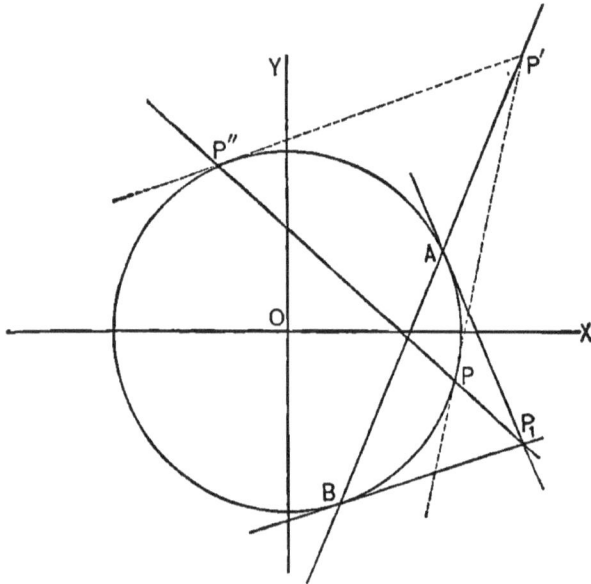

FIG. 27.

which the chord, in any position, cuts the circle; hence (1) will be satisfied for the co-ordinates of the points of intersection of *every* pair of tangents so drawn.

Equation (1) is, therefore, the equation of the required locus. We observe that equation (1) is identical with (1) of the preceding article; hence the chord PP'' is the locus whose equation we sought.

The point P' (x', y') is called THE POLE of the line PP'' $\left(\dfrac{x'x}{a^2} + \dfrac{y'y}{a^2} = 1\right)$, and the line PP'' $\left(\dfrac{x'x}{a^2} + \dfrac{y'y}{a^2} = 1\right)$ is called

THE POLAR of the point P' (x', y') with regard to the circle

$$\frac{x^2}{a^2} + \frac{y^2}{a^2} = 1.$$

As the principles here developed are perfectly general, the pole may be *without*, *on*, or *within* the circle.

Let the student prove that the line joining the pole and the centre is perpendicular to the polar.

NOTE. — The terms *pole* and *polar* used in this article have no connection with the same terms used in treating of polar co-ordinates, Chapter I.

50. *If the polar of the point P' (x', y'), Fig. 27, passes through P_1 (x_1, y_1), then the polar of P_1 (x_1, y_1) will pass through P' (x', y').*

The equation of the polar to P' (x', y') is

$$\frac{x'x}{a^2} + \frac{y'y}{a^2} = 1.$$

In order that P_1 (x_1, y_1) may be on this line, we must have,

$$\frac{x'x_1}{a^2} + \frac{y'y_1}{a^2} = 1.$$

But this is also the equation of condition that the point P' (x', y') may lie on the line whose equation is

$$\frac{x_1 x}{a^2} + \frac{y_1 y}{a^2} = 1.$$

But this is the equation of the polar of P_1 (x_1, y_1); hence the proposition.

51. *To ascertain the relationship between the conjugate diameters of the circle.*

A pair of diameters are said to be conjugate when they are so related that when the curve is referred to them as axes its equation will contain only the second powers of the variables.

Let $\qquad x^2 + y^2 = a^2 \ldots$ (1)

be the equation of the circle, referred to its centre and axes. To ascertain what this equation becomes when referred to OY', OX', axes making any angle with each other, we must substitute in the rectangular equation the values of the old

co-ordinates in terms of the new.   From Art. 33, Cor. 1, we
have

$$x = x' \cos \theta + y' \cos \varphi$$
$$y = x' \sin \theta + y' \sin \varphi$$

for the equations of transformation.   Substituting these
values in (1) and reducing, we have,

$$y'^2 + 2 x'y' \cos (\varphi - \theta) + x'^2 = a^2 \ . \ . \ . \ (2)$$

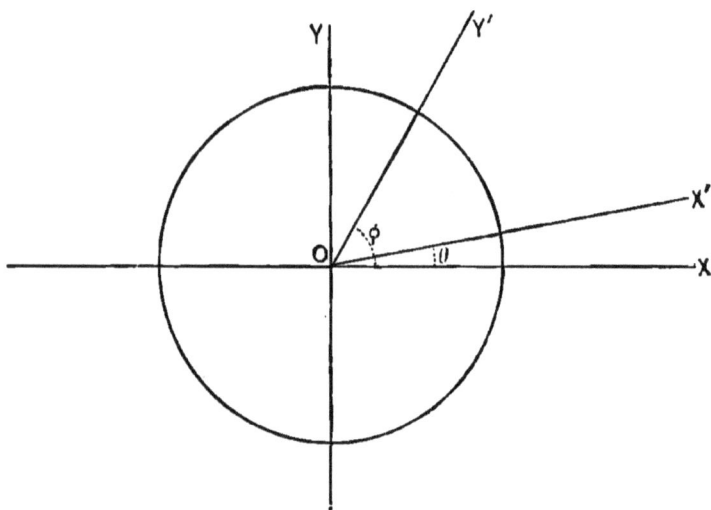

FIG. 28.

Now, in order that OY', OX' may be conjugate diameters
they must be so related that the term containing $x'y'$ in (2)
must disappear; hence the equation of condition,

$$\cos (\varphi - \theta) = 0 \, ;$$
$$\therefore \varphi - \theta = 90°, \text{ or } \varphi - \theta = 270°.$$

The conjugate diameters of the circle are therefore perpendicular to each other.   As there are an infinite number of
pairs of lines in the circle which satisfy the condition of being
at right angles to each other, it follows that *in the circle there*
are an infinite number of conjugate diameters.

EXAMPLES.

1. Prove that the line $y = \sqrt{3}\,x + 10$ touches the circle $x^2 + y^2 = 25$, and find the co-ordinates of the point of tangency.

Ans. Point of tangency $\left(-\dfrac{5}{2}\sqrt{3},\ \dfrac{5}{2} \right)$.

2. What must be the value of b in order that the line $y = 2x + b$ may touch the circle $x^2 + y^2 = 16$?

Ans. $b = \pm \sqrt{80}$.

3. What must be the value of s in order that the line $y = sx - 4$ may touch the circle $x^2 + y^2 = 2$?

Ans. $s = \pm \sqrt{7}$.

4. The slope of a pair of parallel tangents to the circle $x^2 + y^2 = 16$ is 2; required their equations.

Ans. $\begin{cases} y = 2x + \sqrt{80}. \\ y = 2x - \sqrt{80}. \end{cases}$

Two tangents are drawn from a point to a circle; required the equation of the chord joining the points of tangency in each of the following cases:

5. From $(4, 2)$ to $x^2 + y^2 = 9$.

Ans. $4x + 2y = 9$.

6. From $(3, 4)$ to $x^2 + y^2 = 8$.

Ans. $3x + 4y = 8$.

7. From $(1, 5)$ to $x^2 + y^2 = 16$.

Ans. $x + 5y = 16$.

8. From (a, b) to $x^2 + y^2 = c^2$.

Ans. $ax + by = c^2$.

What are the equations of the polars of the following points:

9. Of $(2, 5)$ with regard to the circle $x^2 + y^2 = 16$?

Ans. $\dfrac{2x}{16} + \dfrac{5y}{16} = 1$.

10. Of $(3, 4)$ with regard to the circle $x^2 + y^2 = 9$?

Ans. $3x + 4y = 9$.

11. Of (a, b) with regard to the circle $x^2 + y^2 = m$?

Ans. $ax + by = m$.

What are the poles of the following lines:

12. Of $2x + 3y = 5$ with regard to the circle $x^2 + y^2 = 25$?

Ans. $(10, 15)$.

13. Of $\dfrac{x}{2} + y = 4$ with regard to the circle

$$\frac{x^2}{16} + \frac{y^2}{16} = 1 ?$$ Ans. $(2, 4)$.

14. Of $y = sx + b$ with regard to the circle

$$\frac{x^2}{a^2} + \frac{y^2}{a^2} = 1 ?$$

Ans. $\left(-\dfrac{a^2 s}{b}, \dfrac{a^2}{b} \right)$.

15. Find the equation of a straight line passing through $(0, 0)$ and touching the circle $x^2 + y^2 - 3x + 4y = 0$.

Ans. $y = \dfrac{3}{4}x$.

GENERAL EXAMPLES.

1. Find the equation of that diameter of a circle which bisects all chords drawn parallel to $y = sx + b$.

Ans. $sy + x = 0$.

2. Required the co-ordinates of the points in which the line $2y - x + 1 = 0$ intersects the circle

$$\frac{x^2}{4} + \frac{y^2}{4} = 1.$$

3. Find the co-ordinates of the points in which two lines drawn through $(3, 4)$ touch the circle

$$\frac{x^2}{9} + \frac{y^2}{9} = 1.$$

[The points are common to the chord of contact and the circle.]

4. The centre of a circle which touches the Y-axis is at $(4, 0)$; required its equation.

Ans. $(x - 4)^2 + y^2 = 16.$

5. Find the equation of the circle whose centre is at the origin and to which the line $y = x + 3$ is tangent.

Ans. $2 x^2 + 2 y^2 = 9.$

6. Given $x^2 + y^2 = 16$ and $(x - 5)^2 + y^2 = 4$; required the equation of the circle which has their common chord for a diameter.

7. Required the equation of the circle which has the distance of the point $(3, 4)$ from the origin as its diameter.

Ans. $x^2 + y^2 - 3 x - 4 y = 0.$

8. Find the equation of the circle which touches the lines represented by $x = 3$, $y = 0$, and $y = x$.

9. Find the equation of the circle which passes through the points $(1, 2)$, $(-2, 3)$, $(-1, -1)$.

10. Required the equation of the circle which circumscribes the triangle whose sides are represented by $y = 0$, $3 y = 4 x$, and $3 y = -4 x + 6$.

Ans. $x^2 + y^2 - \frac{6}{4} x - \frac{33}{16} y = 0.$

11. Required the equation of the circle whose intercepts are a and b, and which passes through the origin.

Ans. $x^2 + y^2 - ax - by = 0.$

12. The points $(1, 5)$ and $(4, 6)$ lie on a circle whose centre is in the line $y = x - 4$; required its equation.

Ans. $2 x^2 + 2 y^2 - 17 x - y = 30.$

13. The point $(3, 2)$ is the middle point of a chord of the circle $x^2 + y^2 = 16$; required the equation of the chord.

14. Given $x^2 + y^2 = 16$ and the chord $y - 4 x = 8$. Show that a perpendicular from the centre of the circle bisects the chord.

15. Find the locus of the centres of all the circles which pass through $(2, 4)$, $(3, -2)$.

16. Show that if the polars of two points meet in a third point, then that point is the pole of the line joining the first two points.

17. Required the equation of the circle whose sub-tangent $= 8$, and whose sub-normal $= 2$.

$$Ans. \quad x^2 + y^2 = 20.$$

18. Required the equation of the circle whose sub-normal $= 2$, the distance of the point in which the tangent intersects the X-axis from the origin being $= 8$.

$$Ans. \quad x^2 + y^2 = 16.$$

19. Required the conditions in order that the circles $ax^2 + ay^2 + cx + dy + e = 0$ and $ax^2 + ay^2 + kx + ly + m = 0$ may be concentric.

$$Ans. \quad c = k, \ d = l.$$

20. Required the polar co-ordinates of the centre and the radius of the circle

$$r^2 - 2\,r\,(\cos \theta + \sqrt{3} \sin \theta) = 5.$$

$$Ans. \quad (2, 60°)\,; \ r = 3.$$

21. A line of fixed length so moves that its extremities remain in the co-ordinate axes; required the equation of the circle generated by its middle point.

22. Find the locus of the vertex of a triangle having given the base $= 2\,a$ and the sum of the squares of its sides $= 2\,b^2$.

$$Ans. \quad x^2 + y^2 = b^2 - a^2.$$

23. Find the locus of the vertex of a triangle having given the base $= 2\,a$ and the ratio of its sides

$$= \frac{m}{n}.$$

$$Ans. \quad \text{A circle.}$$

24. Find the locus of the middle points of chords drawn from the extremity of any diameter of the circle

$$\frac{x^2}{a^2} + \frac{y^2}{a^2} = 1.$$

CHAPTER VI.

THE PARABOLA.

52. THE parabola is the locus generated by a point moving in the same plane so as to remain always equidistant from a fixed point and a fixed line.

The fixed point is called the Focus; the fixed line is called the DIRECTRIX; the line drawn through the focus perpendicular to the directrix is called the AXIS; the point on the axis midway between the focus and directrix is called the VERTEX of the parabola.

53. *To find the equation of the parabola, given the focus and directrix.*

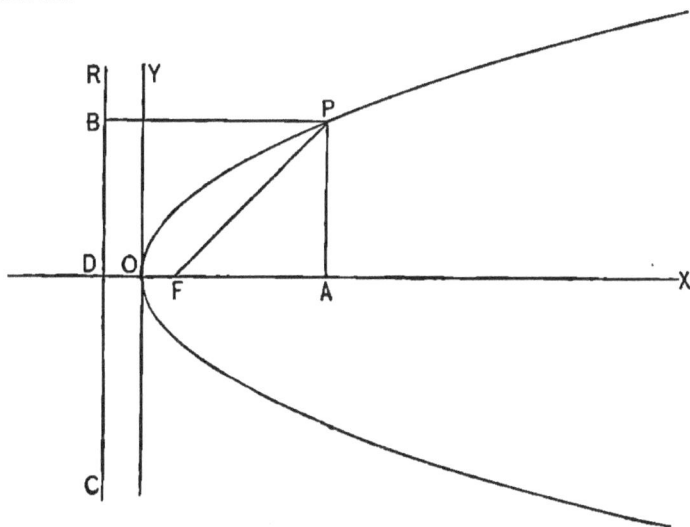

FIG. 29.

Let RC be the directrix and let F be the focus. Let OX, the axis of the curve, and the tangent OY drawn at the vertex

O, be the co-ordinate axes.   Take any point P on the curve and draw PA ∥ to OY, PB ∥ to OX and join P and F.   Then

$$(OA, AP) = (x, y)$$ are the co-ordinates of P.

From the right angled triangle FAP, we have

$$y^2 = AP^2 = FP^2 - FA^2; \ \ . \ . \ . \ (1)$$

But from the mode of generating the curve, we have

$$FP^2 = BP^2 = (AO + OD)^2 = (x + OD)^2,$$

and from the figure, we have

$$FA^2 = (AO - OF)^2 = (x - OF)^2.$$

Substituting these values in (1), we have

$$y^2 = (x + OD)^2 - (x - OF)^2. \ . \ . \ (2)$$

Let $DF = p$, then $OD = OF = \dfrac{p}{2}$;   hence

$$y^2 = \left(x + \frac{p}{2}\right)^2 - \left(x - \frac{p}{2}\right)^2;$$

or, after reduction, $y^2 = 2\,px \ . \ . \ . \ (3)$

As equation (3) is true for *any* point of the parabola it is true for every point; hence it is the equation of the curve.

Cor. 1.   If (x', y') and (x'', y'') are the co-ordinates of any two points on the parabola, we have,

$$y'^2 = 2\,px' \text{ and } y''^2 = 2\,px'';$$

hence $y'^2 : y''^2 :: x' : x''$;

i.e.. *the squares of the ordinates of any two points on the parabola are to each other as their abscissas.*

Schol.   By interchanging x and y, or changing the sign of the second member, or both in (3), we have

$y^2 = -2\,px$ for the equation of a parabola symmetrical with respect to X and extending to the left of Y;

$x^2 = 2\,py$ for the equation of a parabola symmetrical with respect to Y and extending above X.   .

$x^2 = -2\,py$ for the equation of a parabola symmetrical with respect to Y and extending below X.

Let the student discuss each of these equations.   See Art. 13.

54. *To construct the parabola, given the focus and directrix.*

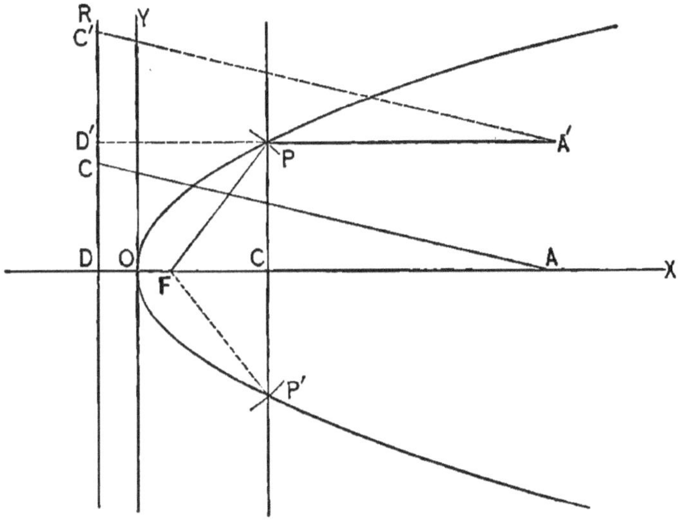

FIG. 30.

First Method. — Let DR be the directrix and let F be the focus.

From F let fall the perpendicular FD on the directrix; it will be the axis of the curve. Take a triangular ruler ADC and make its base and altitude coincide with the axis and directrix, respectively. Attach one end of a string, whose length is AD, to A; the other end to a pin fixed at F. Place the point of a pencil in the loop formed by the string and stretch it, keeping the point of the pencil pressed against the base of the triangle. Now, sliding the triangle up a straight edge placed along the directrix, the point of the pencil will describe the arc OP of the parabola; for in every position of the pencil point the condition of its being equally distant from the focus and directrix is satisfied. It is easily seen, for instance, that when the triangle is in the position A'D'C' that FP = PD'.

Second Method. — Take any point C on the axis and erect

the perpendicular P'CP. Measure the distance DC. With F as a centre and DC (= FP') as a radius describe the arc of a circle, cutting P'CP in P and P'. P and P' will be points of the parabola. By taking other points along the axis we may, by this method, locate as many points of the curve as may be desired.

55. *To find the Latus-rectum, or parameter of the parabola.* The LATUS-RECTUM, *or* PARAMETER *of the parabola, is the double ordinate passing through the focus.*

The abscissa of the points in which the latus-rectum pierces the parabola is $x = \dfrac{p}{2}$.

Making this substitution in the equation
$$y^2 = 2px$$
we have $y^2 = 2p\dfrac{p}{2} = p^2.$

Hence $2y = 2p.$

COR. 1. Forming a proportion from the equation
$$y^2 = 2px,$$
we have $x : y :: y : 2p;$

i.e., *the latus-rectum of the parabola is a third proportional to any abscissa and its corresponding ordinate.*

EXAMPLES.

Find the latus-rectum and write the equation of the parabola which contains the point:

1. $(2, 4)$.
   Ans. $8, y^2 = 8x.$

3. (a, b).
   Ans. $\dfrac{b^2}{a}, y^2 = \dfrac{b^2}{a}x.$

2. $(-2, 4)$.
   Ans. $-8, y^2 = -8x.$

4. $(-a, 2)$.
   Ans. $-\dfrac{4}{a}, y^2 = -\dfrac{4}{a}x.$

5. What is the latus-rectum of the parabola $x^2 = 2py$? How is it defined in this case?

6. What is the equation of the line which passes through the vertex and the positive extremity of the latus-rectum of any parabola whose equation is of the form $y^2 = 2\,px$?

Ans. $y = 2\,x.$

7. The focus of a parabola is at 2 units' distance from the vertex of the curve; what is its equation

(a) when symmetrical with respect to the X-axis ?

(b)   "   "   "   "   "   "   Y-axis ?

Ans. (a) $y^2 = 8\,x,$ (b) $x^2 = 8\,y.$

Construct each of the following parabolas by three different methods.

8. $y^2 = 8\,x.$ **10.** $x^2 = 6\,y.$

9. $y^2 = -4\,x.$ **11.** $x^2 = -10\,y.$

12. What are the co-ordinates of the points on the parabola $y^2 = 6\,x$ where the ordinate and abscissa are equal ?

Ans. (0, 0), and (6, 6).

13. Required the co-ordinates of the point on the parabola $x^2 = 4\,y$ whose ordinate and abscissa bear to each other the ration $3:2.$ *Ans.* (6, 9).

14. What is the equation of the parabola when referred to the directrix and X-axis as axes ?   *Ans.* $y^2 = 2\,px - p^2.$

Find the points of intersection of the following:

15. $y^2 = 4\,x$ and $2\,y - x = 0.$

Ans. (0, 0), (16, 8).

16. $x^2 = 6\,y$ and $y - x - 1 = 0.$

17. $y^2 = -8\,x$ and $x + 3 = 0.$

18. $y^2 = 2\,x$ and $x^2 + y^2 = 8.$

Ans. (2, 2), (2, -2).

19. $x^2 = -4\,y$ and $3\,x^2 + 2\,y^2 = 6.$

20. $x^2 = 4\,y$ and $y^2 = 4\,x.$

56. *To deduce the polar equation of the parabola, the focus being taken as the pole.*

The equation of the parabola referred to OY, OX, Fig. 29, is

$$y^2 = 2\,px \ \ldots \ (1)$$

To refer the curve to the initial line FX and the pole F $\left(\dfrac{p}{2},\, 0\right)$ we have for the equations of transformation, Art. 34, Cor. 1,

$$x = \frac{p}{2} + r \cos \theta.$$

$$y = r \sin \theta.$$

Substituting these values in (1), we have

$$r^2 \sin^2 \theta = p^2 + 2\,pr \cos \theta.$$

But $\sin^2 \theta = 1 - \cos^2 \theta$;

$$\therefore r^2 = p^2 + 2\,pr \cos \theta + r^2 \cos^2 \theta = (p + r \cos \theta)^2,$$
$$\therefore r = p + r \cos \theta,$$

or, solving,

$$r = \frac{p}{1 - \cos \theta} \ \ldots \ (2)$$

is the required equation.

We might have deduced this value directly as follows :
Let P (r, θ) Fig. 29 be any point on the curve; then

$$FP = DA = DF + FA = p + r \cos \theta;$$

i. e., $r = p + r \cos \theta.$

Hence $r = \dfrac{p}{1 - \cos \theta}.$

Cor. 1. If $\theta = 0$, $r = \infty$.

If $\theta = 90°$, $r = p.$

If $\theta = 180°$, $r = \dfrac{p}{2}.$

If $\theta = 270°$, $r = p.$

If $\theta = 360°$, $r = \infty.$

An inspection of the figure will verify these results.

57. *To deduce the equation of the tangent to the parabola.*

If (x', y'), (x'', y'') be the points in which a secant line cuts the parabola, then

$$y - y' = \frac{y' - y''}{x' - x''} (x - x') \quad \ldots (1)$$

will be its equation. Since (x', y'), (x'', y'') are points of the parabola, we have

$$y'^2 = 2 px' \quad \ldots (2)$$
$$y''^2 = 2 px'' \quad \ldots (3)$$

These three equations must subsist at the same time; hence, subtracting (3) from (2) and factoring, we have

$$(y' - y'')(y' + y'') = 2 p (x' - x'');$$

i.e.,
$$\frac{y' - y''}{x' - x''} = \frac{2 p}{y' + y''}.$$

Substituting this value in (1), the equation of the secant becomes

$$y - y' = \frac{2 p}{y' + y''} (x - x') \quad \ldots (4)$$

When the secant, revolved about (x'', y''), becomes tangent to the parabola (x', y') coincides with (x'', y''); hence $x' = x''$, $y' = y''$. Making this substitution in (4), we have,

$$y - y'' = \frac{p}{y''} (x - x''). \quad \ldots (5)$$

or, simplifying, recollecting that $y''^2 = 2 px''$, we have

$$yy'' = p (x + x'') \quad \ldots (6)$$

for the *equation of the tangent to the parabola.*

58. *To deduce the value of the sub-tangent.*

Making $y = 0$ in (6), Art. 57, we have

$$x = - x'' = OT, \quad \text{(Fig. 31)}$$

for the abscissa of the point in which the tangent intersects the X-axis. But the sub-tangent CT is the distance of this point from the foot of the ordinate of the point of tangency; i.e., twice the distance just found; hence

$$\textit{Sub-tangent} = 2 x'';$$

i.e., *the sub-tangent is equal to double the abscissa of the point of tangency.*

59. The preceding principle affords us a simple method of constructing a tangent to a parabola at a given point.

Let P″ (x'', y'') be any point of the curve. Draw the ordinate P″C, and measure OC. Lay off OT = OC.

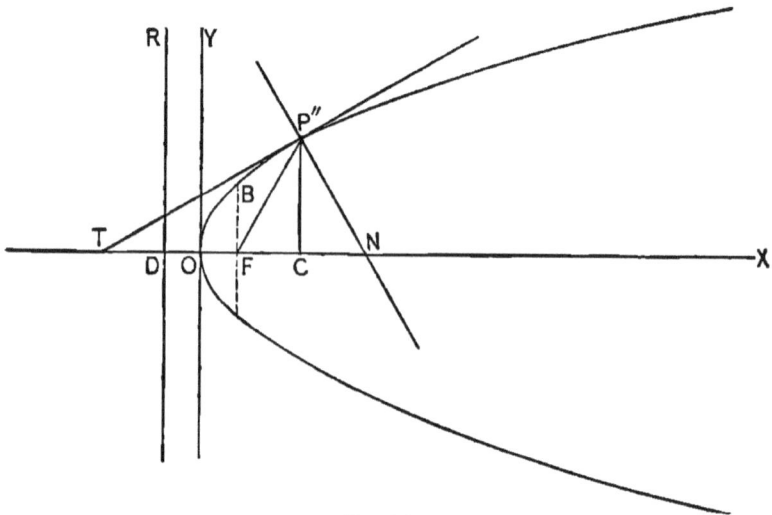

Fig. 31.

A line joining T and P″ will be tangent to the parabola at P″.

60. *To deduce the equation of the normal to the parabola.*

The equation of any line through P″ (x'', y'') Fig. 31, is

$$y - y'' = s (x - x'') \ \ldots (1)$$

We have found Art. 57, (5) for the slope of the tangent P″T

$$s' = \frac{p}{y''};$$

hence, for the slope of the normal P″N, we have

$$s = - \frac{y''}{p}.$$

Substituting this value of s in (1), we have

$$y - y'' = -\frac{y''}{p}(x - x'') \quad \cdots \quad (2)$$

for *the equation of the normal to the parabola.*

61. *To deduce the value of the sub-normal.*

Making $y = 0$ in (2) Art. 60, we have, after reduction,

$$x = p + x'' = ON; \text{ Fig. 31,}$$
$$\therefore \textit{Sub-normal} = NC = p + x'' - x'' = p.$$

Hence the sub-normal in the parabola is constant and equal to the semi-parameter FB.

62. *To show that the tangents drawn at the extremities of the latus rectum are perpendicular to each other.*

The co-ordinates of the extremities of the latus-rectum are $\left(\frac{p}{2}, p\right)$ for the upper point, and $\left(\frac{p}{2}, -p\right)$ for the lower point.

Substituting these values successively in the general equation of the tangent line, Art. 57 (6), we have

$$yp = p\left(x + \frac{p}{2}\right),$$

$$-yp = p\left(x + \frac{p}{2}\right),$$

or, cancelling,

$$y = x + \frac{p}{2} \quad \cdots \quad (1)$$

$$y = -x - \frac{p}{2} \quad \cdots \quad (2)$$

for the equations of the tangents. As the coefficient of x in (2) is minus the reciprocal of the coefficient of x in (1), the lines are perpendicular to each other.

Cor. 1. Making $y = 0$ in (1) and (2), we find in each case that $x = -\frac{p}{2}$; hence, *the tangents at the extremities of the*

*latus-rectum and the directrix meet the axis of the parabola
in the same point.*

The values of the coefficients of x in (1) and (2) show that
these tangent lines make angles of 45° with the X-axis.

63. *To deduce the equation of the parabola when referred to
the tangents at the extremities of the latus-rectum as axes.*

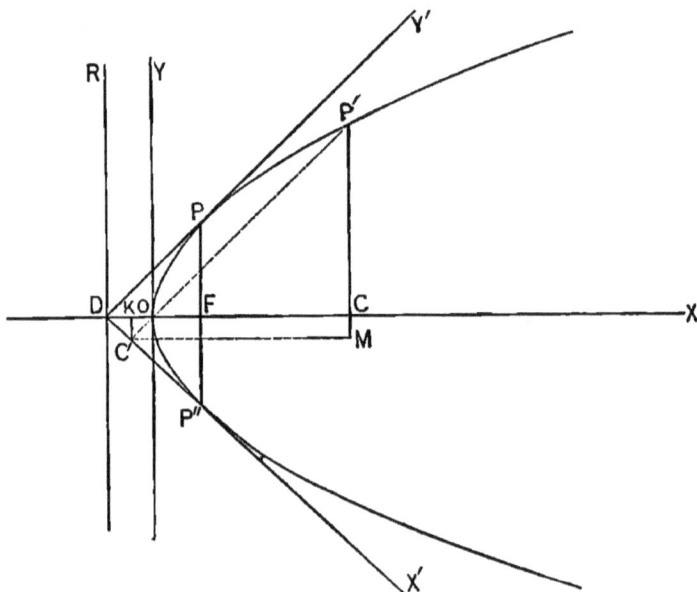

FIG. 32.

The equation of the parabola when referred to OY, OX, is

$$y^2 = 2\,px \ . \ . \ . \ (1).$$

We wish to ascertain what this equation becomes when the
curve is referred to DY′, DX′. as axes.

Let P′ (x', y') be any point of the curve; then, Fig. 32

$$(OC,\ CP'') = (x,\ y),\ \text{and}\ (DC',\ C'P'') = (x',\ y').$$

From the figure, we have,

$$OC = DC - DO = DK + C'M - DO;$$

but

$$\mathrm{DK} = x' \cos 45° = \frac{x'}{\sqrt{2}}, \; \mathrm{C'M} = y' \cos 45° = \frac{y'}{\sqrt{2}}, \; \mathrm{DO} = \frac{p}{2};$$

hence $\qquad x = \dfrac{x'}{\sqrt{2}} + \dfrac{y'}{\sqrt{2}} - \dfrac{p}{2}.$

We have, also, $\quad \mathrm{CP'} = \mathrm{MP'} - \mathrm{C'K};$

i.e., $\qquad\qquad y = \dfrac{y'}{\sqrt{2}} - \dfrac{x'}{\sqrt{2}}.$

Substituting the values of x and y in (1), we have,

$$\frac{1}{2}(y' - x')^2 = \frac{2p}{\sqrt{2}}(x' + y') - p^2 \cdots (2)$$

In order to simplify this expression let $\mathrm{DP} = a$; then from the triangle DPF, we have,

$$\mathrm{DF} = p = a \cos 45° = \frac{a}{\sqrt{2}}.$$

Substituting this value of p in (2) and multiplying through by 2, we have, $(y' - x')^2 = 2a(x' + y') - a^2,$

or, $\qquad y'^2 + x'^2 - 2x'y' - 2ax' - 2ay' + a^2 = 0.$

Adding $4x'y'$ to both members, the equation takes the form

$$(x' + y' - a)^2 = 4x'y',$$

or $\qquad\qquad x' + y' - a = \pm 2x'^{\frac{1}{2}} y'^{\frac{1}{2}};$

∴ transposing, $\; x' \pm 2x'^{\frac{1}{2}} y'^{\frac{1}{2}} + y' = a;$

$$\therefore x'^{\frac{1}{2}} \pm y'^{\frac{1}{2}} = \pm a^{\frac{1}{2}}, \cdots (3)$$

or, symmetrically, dropping accents,

$$\frac{x^{\frac{1}{2}}}{a^{\frac{1}{2}}} \pm \frac{y^{\frac{1}{2}}}{a^{\frac{1}{2}}} = \pm 1 \cdots (4)$$

is the required equation.

EXAMPLES.

1. What is the polar equation of the parabola, the pole being taken at the vertex of the curve ?

Ans. $r = 2\,p \cot \theta \csc \theta.$

Find the equation of the tangent to each of the following parabolas, and give the value of the subtangent in each case :

2. $y^2 = 4\,x$ at $(1, 2)$.   *Ans.* $y = x + 1$; 2.

3. $x^2 = 4\,y$ at $(- 2, 1)$.   *Ans.* $x + y + 1 = 0$; 2.

4. $y^2 = - 6\,x$ at $(- 6, \text{ord} +)$.   *Ans.* $2\,y + x = 6$; 12.

5. $x^2 = - 8\,y$ at $(\text{abs} +, - 2)$.   *Ans.* $x + y = 2$; 4.

6. $y^2 = 4\,ax$ at $(a, - 2\,a)$.

7. $y^2 = mx$ at (m, m).

8. $x^2 = - py$ at $(\text{abs} +, - p)$.

9. $x^2 = 2\,py$ at $\left(abs -, \dfrac{p}{8}\right).$

Write the equation of the normal to each of the following parabolas :

10. To $y^2 = 16\,x$ at $(1, 4)$.

11. To $x^2 = - 10\,y$ at $(\text{abs} +, - 2)$.

12. To $y^2 = - mx$ at $(- m, m)$.

13. To $x^2 = 2\,my$ at $\left(abs -, \dfrac{m}{8}\right).$

14. The equation of a parabola is $x^{\frac{1}{2}} \pm y^{\frac{1}{2}} = \pm a^{\frac{1}{2}}$; what are the co-ordinates of the vertex of the curve ?

Ans. $\left(\dfrac{1}{4}\,a, \dfrac{1}{4}\,a\right).$

15. Given the parabola $y^2 = 4\,x$ and the line $y - x = 0$; required the equation of the tangent which is,

(*a*) parallel to the line,

(*b*) perpendicular to the line.

Ans. (*a*) $y = x + 1$, (*b*) $y + x + 1 = 0$.

16. The point $(-1, 2)$ lies outside the parabola $y^2 = 6x$; what are the equations of the tangents through the point to the parabola?

17. The point $(2, 45°)$ is on a parabola which is symmetrical with respect to the X-axis; required the equation of the parabola, the pole being at the focus.

$$Ans. \quad y^2 = (4 - 2\sqrt{2})\,x.$$

18. The subtangent of a parabola $= 10$ for the point $(5, 4)$; required the equation of the curve and the value of the subnormal.

$$Ans. \quad y^2 = \frac{16}{5}x\,;\ \frac{8}{5}.$$

64. *The tangent to the parabola makes equal angles with the focal line drawn to the point of tangency and the axis of the curve.*

From Fig. 31 we have,

$$FT = FO + OT = \frac{p}{2} + x''.$$

We have, also,

$$FP'' = DC = DO + OC = \frac{p}{2} + x''.$$

$$\therefore FT = FP''.$$

The triangle FP'''T is therefore isosceles and

$$FP''T = FTP''.$$

65. *To find the condition that the line $y = sx + c$ must fulfil in order to touch the parabola $y^2 = 2\,px$.*

Eliminating y from the two equations, and solving the resulting equation with respect to x, we have,

$$x = \frac{p - sc \pm \sqrt{(cs - p)^2 - c^2 s^2}}{s^2} \quad \dots \quad (1)$$

for the abscissæ of the points of intersection of the parabola and line, considered as a secant. When the secant becomes a tangent, these abscissas become equal; but the condition for equality of abscissas is that the radical in the numerator of (1) shall be zero; hence

$$(cs - p)^2 - c^2 s^2 = 0,$$

or, solving, $\quad c = \dfrac{p}{2\,s}$

is the condition that the line must fulfil in order to touch the parabola.

Cor. 1. Substituting the value of c in the equation

$$y = sx + c,$$

we have, $\quad y = sx + \dfrac{p}{2\,s} \; \ldots \; (2)$

for *the equation of the tangent in terms of its slope.*

66. *To find the locus generated by the intersection of a tangent, and a perpendicular to it from the focus as the point of tangency moves around the curve.*

The equation of a straight line through the focus $\left(\dfrac{p}{2}, 0 \right)$ is

$$y = s' \left(x - \dfrac{p}{2} \right) \; \ldots \; (1)$$

In order that this line shall be perpendicular to the tangent

$$y = sx + \dfrac{p}{2\,s} \; \ldots \; (2)$$

we must have, $\quad s' = -\dfrac{1}{s};$

hence $\quad y = -\dfrac{1}{s} x + \dfrac{p}{2\,s} \; \ldots \; (3)$

is the equation of a line through the focus perpendicular to the tangent. Subtracting (3) from (2), we have

$$\left(s + \dfrac{1}{s} \right) x = 0,$$

or, $\quad x = 0,$

for the equation of the required locus. But $x = 0$ is the equation of the Y-axis; hence, *the perpendiculars from the focus to the tangents of a parabola intersect the tangents on the* Y-*axis.*

67. *To find the locus generated by the intersection of two tangents which are perpendicular to each other as the points of tangency moves around the curve.*

The equation of a tangent to the parabola is, Art. 65 (2),

$$y = sx + \frac{p}{2s} \cdots (1)$$

The equation of a perpendicular tangent is

$$y = -\frac{1}{s}x - \frac{ps}{2} \cdots (2)$$

Subtracting (2) from (1), we have,

$$\left(s + \frac{1}{s}\right)x + \left(s + \frac{1}{s}\right)\frac{p}{2} = 0;$$

$$\therefore x = -\frac{p}{2} \cdots (3)$$

is the equation of the required locus. But (3) is the equation of the directrix; hence, *the intersection of all perpendicular tangents drawn to the parabola are points of the directrix.*

68. *Two tangents are drawn to the parabola from a point without; required the equation of the line joining the points of tangency.*

Let (x', y') be the given point without the parabola, and let (x'', y''), (x_2, y_2) be the points of tangency. Since (x', y') is on both tangents, its co-ordinates must satisfy their equations; hence, the equations of condition,

$$y'y'' = p(x' + x''),$$
$$y'y_2 = p(x' + x_2).$$

The two points of tangency (x'', y''), (x_2, y_2) must therefore satisfy

$$y'y = p(x' + x),$$

or $\qquad y y' = p(x + x') \cdots (1)$

Since (1) is the equation of a straight line, and is satisfied for the co-ordinates of both points of tangency, it is the equation of the line joining those points.

69. *To find the equation of the polar of the pole* ($x,'$ y') *with regard to the parabola* $y^2 = 2\,px$.

The polar of a pole with regard to a given curve is the line generated by the point of intersection of a pair of tangents drawn to the curve at the points in which a secant line through the pole intersects the curve as the secant line revolves about the pole.

By a course of reasoning similar to that of **Art. 49**, we may prove the required equation to be

$$yy' = p\,(x + x') \;\ldots\; (1)$$

As the reasoning by means of which (1) is deduced is perfectly general, the pole may be *without*, *on*, or *within* the parabola.

Cor. 1. If we make, in (1), (x', y') $= \left(\dfrac{p}{2},\ 0\right)$, we have

$$x = -\frac{p}{2};$$

hence, *the directrix is the polar of the focus.*

70. *To ascertain the position and direction of the axes, other than the axis of the parabola and the tangent at the vertex, to which if the parabola be referred its equation will remain unchanged in form.*

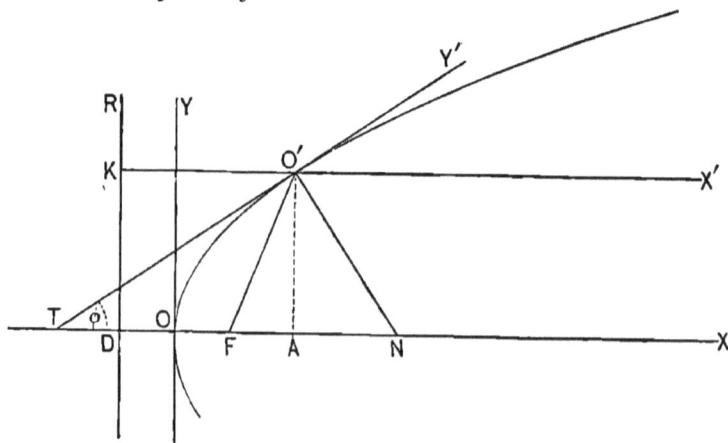

FIG. 33.

Since the equation is to retain the form

$$y^2 = 2\,px \ldots (1)$$

let
$$y'^2 = 2\,p'x' \ldots (2)$$

be the equation of the parabola when referred to the axes, whose position and direction we are now seeking. It is obvious at the outset that whatever may be the position of the axes relatively to each other, the new Y'-axis must be *tangent* to the curve, and the new origin must be *on* the curve; for, if in (2) we make $x' = 0$, we have $y' = \pm\,0$, a result which we can only account for by assuming the Y'-axis and the new origin in the positions indicated. This conclusion, we shall see, is fully verified by the analysis which follows.

Let us refer the curve to a pair of oblique axes, making any angle with each other, the origin being anywhere in the plane of the curve. The equations of transformation are, Art. 33 (1),

$$x = a + x'\cos\theta + y'\cos\varphi$$
$$y = b + x'\sin\theta + y'\sin\varphi.$$

Substituting these values in (1), we have,

$$y'^2\sin^2\varphi + 2\,x'y'\sin\theta\sin\varphi + x'^2\sin^2\theta + 2\,(b\sin\varphi - p\cos\varphi)\,y' + 2\,(b\sin\theta - p\cos\theta)\,x' + b^2 - 2\,pa = 0 \ldots (3)$$

Now, in order that this equation shall reduce to the same form as (1), we must have the following conditions satisfied :

     (*a*) $\sin\theta\sin\varphi = 0.$

     (*b*) $\sin^2\theta = 0.$

     (*c*) $b^2 - 2\,pa = 0.$

     (*d*) $b\sin\varphi - p\cos\varphi = 0.$

If $\theta = 0$, then $\sin\theta\sin\varphi = 0$ and $\sin^2\theta = 0$; i.e., conditions (*a*) and (*b*) are satisfied for this assumed value of θ. But θ is the angle which the new X'-axis makes with the old X-axis; hence, these axes are parallel.

If (*a*, *b*) be a point of the parabola $y^2 = 2\,px$. then $b^2 = 2\,pa$ is an analytical expression of the fact; hence (*c*) shows that the new origin lies on the curve.

If $\dfrac{\sin\varphi}{\cos\varphi} = \tan\ \varphi = \dfrac{p}{b}$, then (d) is satisfied. But $\dfrac{p}{b}$ is the

slope of the tangent at the point whose ordinate is b, Art. 57, (5), and tan φ is the slope of the new Y'-axis; hence, the new Y'-axis is a tangent to the parabola at the point whose ordinate is b ; $\therefore$ at (a, b) ; $\therefore$ at the new origin.

Cor. 1. Substituting (a), (b), (c), and (d) in (3), recollecting that cos θ = cos 0 = 1, we have, after dropping accents,

$$y^2 = \frac{2p}{\sin^2\varphi}\, x,$$

or, letting $\qquad \dfrac{p}{\sin^2\varphi} = p',$

we have $\qquad y^2 = 2\,p'x \ \ldots\ (4)$

for the equation of the parabola when referred to O'Y', O'X', Fig. 33. The form of (4) shows that for every value assumed for x, y has two values, equal but of opposite sign; hence, *OX' bisects all chords, drawn parallel to O Y' and is therefore a diameter of the parabola.*

Note. — *A* Diameter *of a curve is a line which bisects a system of parallel chords.*

71. *To show that the parameter of any diameter is equal to four times the distance from the focus to the point in which that diameter cuts the curve.*

Draw the focal line FO' and the normal O'N, Fig. 33.

Since the triangle O'FT is isosceles, Art. 64, the angle O'FN = 2 φ.

Since O'N is a normal at O', AO'N = φ and AN = p, Art. 61. Hence in the triangle FO'A

$$AO' = FO' \sin 2\,\varphi = FO'\ 2\sin\varphi\cos\varphi.$$

In the triangle NO'A,

$$AO' = AN \cot\varphi = p\,\frac{\cos\varphi}{\sin\varphi}\,;$$

hence $\qquad FO'\ 2\sin\varphi\cos\varphi = p\,\dfrac{\cos\varphi}{\sin\varphi}\,;$

$$\therefore \mathrm{FO}' = \frac{p}{2\sin^2\varphi}$$

But
$$2\,p' = \frac{2\,p}{\sin^2\varphi};$$
$$\therefore 2\,p' = 4\ \mathrm{FO}'.$$

72. *To find the equation of any diameter in terms of the slope of the tangent and the semi-parameter.*

The equation of any diameter as O'X', Fig. 33, is
$$y = \mathrm{AO}' = b.$$
But from the triangle AO'N, we have,
$$b = \mathrm{AN}\cot\varphi = \frac{p}{\tan\varphi} = \frac{p}{s};$$
hence
$$y = \frac{p}{s} \ \ldots \ (1)$$
is the required equation.

73. *To show that the tangents drawn at the extremities of any chord meet in the diameter which bisects that chord.*

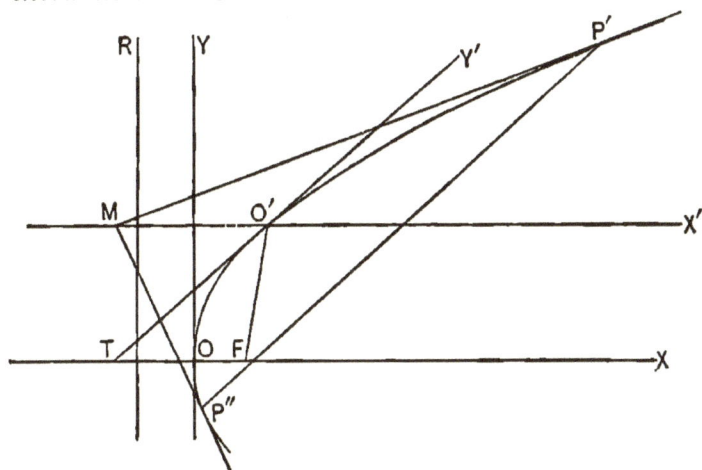

FIG. 34.

Let P' (x', y'), P'' (x'', y'') be the extremities of the chord P'P'';

then
$$y - y' = \frac{y' - y''}{x' - x''}(x - x') \ \ldots \ (1)$$

is its equation. The equation of the tangents at P' (x', y'), P'' (x'', y'') are

$$yy' = p\,(x + x') \ldots (2)$$
$$yy'' = p\,(x + x'') \ldots (3)$$

Eliminating x from (2) and (3) by subtraction, we have,

$$y = p\,\frac{x' - x''}{y' - y''} \ldots (4)$$

for the ordinate of the point of intersection of the tangents.

But $\dfrac{x' - x''}{y' - y''}$ is the reciprocal of the slope of chord P'P'', (see (1)). Hence, since the chord P'P'' and the tangent Y'T are parallel, we have,

$$\frac{x' - x''}{y' - y''} = \frac{1}{s}.$$

Substituting in (4) it becomes

$$y = \frac{p}{s}.$$

Comparing this value of y with (1) of the preceding article, we see that the point of intersection is on the diameter.

EXAMPLES.

1. What must be the value of c in order that the line $y = 4\,x + c$ may touch the parabola $y^2 = 8\,x$?

Ans. ½.

2. What is the parameter of the parabola which the line $y = 3\,x + 2$ touches ?

Ans. 24.

3. The slope of a tangent to the parabola $y^2 = 6\,x$ is $= 3$. What is the equation of the tangent ?

Ans. $y = 3\,x + $ ½.

4. The point (1, 3) lies on a tangent to a parabola ; required the equation of the tangent and the equation of the parabola, the slope of the tangent $= 4$.

Ans. $y = 4\,x - 1$; $y^2 = -16\,x.$

5. In the parabola $y^2 = 8x$ what is the parameter of the diameter whose equation is $y - 16 = 0$?

Ans. 136.

6. Show that if two tangents are drawn to the parabola from any point of the directrix they will meet at right angles.

7. From the point $(-2, 5)$ tangents are drawn to $y^2 = 8x$; required the equation of the chord joining the points of tangency. *Ans.* $5y - 4x + 8 = 0$.

8. What are the equations of the tangents to $y^2 = 6x$ which pass through the point $(-2, 4)$?

Find the equation of the polar of the pole in each of the following cases:

9. Of $(-1, 3)$ with regard to $y^2 = 4x$.

Ans. $3y - 2x + 2 = 0$.

10. Of $(2, 2)$ with regard to $y^2 = -4x$.

Ans. $2y + 2x + 4 = 0$.

11. Of (a, b) with regard to $y^2 = 4x$.

Ans. $by - 2x - 2a = 0$.

12. Given the parabola $y^2 = x$ and the point $(-4, 10)$; to find the intercepts of the polar of the point.

Ans. $a = 4, b = -\dfrac{1}{5}$.

13. The latus-rectum of a parabola $= 4$; required the pole of the line $y - 8x - 4 = 0$.

Ans. $(\tfrac{1}{2}, \tfrac{1}{4})$.

14. Given $y^2 = 10x$ and the tangent $2y - x = 10$; required the equation of the diameter passing through the point of tangency.

Ans. $y = 10$.

GENERAL EXAMPLES.

1. Assuming the equation of the parabola, prove that every point on the curve is equally distant from the focus and directrix.

2. Find the equation of the parabola which contains the points $(0, 0)$, $(2, 3)$, $(-2, 3)$.

Ans. $3\,x^2 = 4\,y$.

3. What are the parameters of the parabolas which pass through the point $(3, 4)$?

Ans. $\tfrac{16}{3}$, and $\tfrac{9}{4}$.

4. Find the equation of that tangent to $y^2 = 9\,x$ which is parallel to the line $y - 2\,x - 4 = 0$.

Ans. $8\,y - 16\,x - 9 = 0$.

5. The parameter of a parabola is 4; required the equation of the tangent line which is perpendicular to the line $y = 2\,x + 2$. Give also the equation of the normal which is parallel to the given line.

6. A tangent to $y^2 = 4\,x$ makes an angle of $45°$ with the X-axis; required the point of tangency.

Ans. $(1, 2)$.

Show that tangents drawn at the extremities of a focal chord

7. Intersect on the directrix.

8. Meet at right angles.

9. That a line joining their point of intersection with the focus is perpendicular to the focal chord.

10. Find the equation of the normal in terms of its slope.

11. Show that from any point within the parabola three normals may be drawn to the curve.

12. Given the parabola $r = \dfrac{4}{1 + \cos \theta}$ to construct the tangent at the point whose vectorial angle $= 60°$, and to find the angle which the tangent makes with the initial line.

Ans. $\theta = 60°$.

13. Find the co-ordinates of the pole, the normal at one extremity of the latus-rectum being its polar.

14. In the parabola $y^2 = 4x$ what is the equation of the chord which the point $(2, 1)$ bisects?

Ans. $y = 2x - 3.$

15. The polar of any point in a diameter is parallel to the ordinates of that diameter.

16. The equation of a chord of $y^2 = 10x$ is $y = 2x - 1$; required the equation of the corresponding diameter.

17. Show that a circle described on a focal chord of the parabola touches the directrix.

18. The base of a triangle $= 2a$ and the sum of the tangents of the base angles $= b$. Show that the locus of the vertex is a parabola.

19. Required the equation of the chord of the parabola $y^2 = 2px$ whose middle point is (m, n).

Ans. $\dfrac{n}{p} = \dfrac{x - m}{y - n}.$

20. A focal chord of the parabola $y^2 = 2px$ makes an angle $= \varphi$ with the X-axis; required its length.

Ans. $\dfrac{2p}{\sin^2 \varphi}.$

21. Show that the focal distance of the point of intersection of two tangents to a parabola is a mean proportional to the focal radii of the points of tangency.

22. Show that the angle between two tangents to a parabola is one-half the angle between the focal radii of the points of tangency.

23. The equation of a diameter of the parabola $y^2 = 2px$ is $y = a$; required the equation of the focal chord which this diameter bisects.

24. The polars of all points on the latus-rectum meet the axis of the parabola $y^2 = 2px$ in the same point; required the co-ordinates of the point.

Ans. $\left(-\dfrac{p}{2}, 0 \right).$

CHAPTER VII

THE ELLIPSE.

74. THE ellipse is the locus of a point so moving in a plane that the *sum* of its distances from two fixed points is always constant and equal to a given line. The fixed points are called the FOCI of the ellipse. If the points are on the given line and equidistant from its extremities, then the given line is called the TRANSVERSE or MAJOR AXIS of the ellipse.

75. *To deduce the equation of the ellipse, given the foci and the transverse axis.*

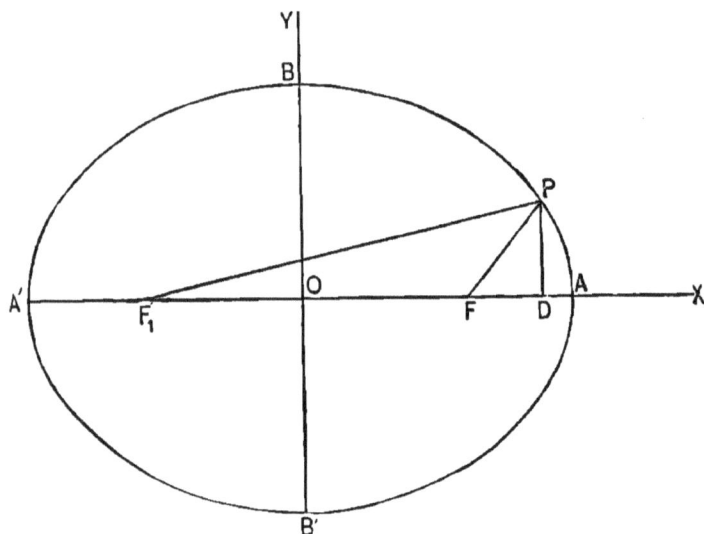

FIG. 35.

Let F, F$_1$ be the foci and AA′ the transverse axis. Draw OY $\perp$ to AA′ at its middle point, and take OY, OX as the co-ordinate axes.

Let P be any point of the curve. Draw PF, PF$_1$; draw also PD ‖ to OY.

Then (OD, DP) = (x, y) are the co-ordinates of P.

Let AA'= $2a$, FF$_1$ = 2OF = 2OF$_1$ = 2 c, FP = r and F$_1$P = r'.

From the right angled triangles FPD and F$_1$PD, we have,

$$r = \sqrt{y^2 + (x - c)^2} \text{ and } r' = \sqrt{y^2 + (x + c)^2} \cdot \cdot \cdot (a)$$

From the mode of generation of the curve, we have,

$$r + r' = 2 a ;$$

hence $\quad \sqrt{y^2 + (x - c)^2} + \sqrt{y^2 + (x + c)^2} = 2 a ; \cdot \cdot \cdot (1)$

or, clearing of radicals, and reducing,

$$a^2 (y^2 + x^2) - c^2 x^2 = a^2 (a^2 - c^2) \cdot \cdot \cdot (2)$$

As this equation (2) expresses the relationship between the co-ordinates of *any* point on the curve, it must express the relationship between the co-ordinates of *every* point; hence it is the required equation.

Equation (2) may be made, however, to assume a more elegant form. Make $x = 0$ in (2), we have,

$$y^2 = a^2 - c^2$$

for the square of the ordinate of the point in which the curve cuts the Y-axis; i.e., $\overline{OB}^2$ ($= \overline{OB}'^2$). Representing this distance by b, we have,

$$b^2 = a^2 - c^2,$$
$$\therefore c^2 = a^2 - b^2 \cdot \cdot \cdot (3)$$

Substituting this value of c^2 in (2) and reducing, we have,

$$a^2 y^2 + b^2 x^2 = a^2 b^2 ; \cdot \cdot \cdot (4)$$

or, symmetrically,

$$\frac{x^2}{a^2} + \frac{y^2}{b^2} = 1 \cdot \cdot \cdot (5)$$

for the equation of ellipse when referred to its centre and axes.

Let the student discuss equation (4). See Art. 12.

Cor. 1. If we make $b = a$ in (4), we have,

$$x^2 + y^2 = a^2$$

which is the equation of a circle.

Cor. 2. If we interchange a and b in (5), we have,

$$\frac{x^2}{b^2} + \frac{y^2}{a^2} = 1 \ . \ . \ . \ (6)$$

for the equation of an ellipse whose transverse axis $(= 2\,a)$ lies along the Y-axis.

Cor. 3. If (x', y') and (x'', y'') are two points on the curve, we have from (4)

$$y'^2 = \frac{b^2}{a^2}\,(a^2 - x'^2) \ and \ y''^2 = \frac{b^2}{a^2}\,(a^2 - x''^2)\,;$$

hence, $y'^2 : y''^2 :: (a - x')\,(a + x') : (a - x'')\,(a + x'')\,;$
i.e., *the squares of the ordinates of any two points on the ellipse are to each other as the rectangles of the segments in which they divide the transverse axis.*

Cor. 4. By making $x = x' - a$ and $y = y'$ in (4), we have after reduction and dropping accents,

$$a^2 y^2 + b^2 x^2 - 2\,ab^2 x = 0 \ . \ . \ . \ (7)$$

for the equation of the ellipse, A' being taken as the origin of co-ordinates.

76. The line BB', Fig. 35, is called the CONJUGATE or MINOR axis of the ellipse; the points A and A' are called the VERTICES of the ellipse. It is evident from the figure that the point O bisects all lines drawn through it and terminating in the curve. For this reason O is called the CENTRE of the ellipse.

The ratio $\dfrac{\sqrt{a^2 - b^2}}{a} = \dfrac{c}{a} = e.$ See (3) Art. 75 . . . (1)

is called the ECCENTRICITY of the ellipse. It is evident that this ratio is always < 1. The value of $c = \pm \sqrt{a^2 - b^2}$ measures the distances of the foci F, F_1 from the centre.

If $a = b$ in (1), then $e = 0$; i.e., when the ellipse becomes a circle its eccentricity becomes zero.

If $b = 0$ in (1), then $e = 1$; i.e., when the ellipse becomes a straight line the eccentricity becomes unity.

77. *To find the values of the focal radii, r, r', of a point on the ellipse in terms of the abscissa of the point.*

The FOCAL RADIUS *of a point on the ellipse is the distance of the point from either focus.*

From equations (a), Art. 75, we have,

$$r = \sqrt{y^2 + (x - c)^2};$$

from the equation of the ellipse, Art. 75 (4), we have,

$$y^2 = \frac{b^2}{a^2}(a^2 - x^2) = b^2 - \frac{b^2}{a^2}x^2;$$

hence, substituting

$$r = \sqrt{b^2 - \frac{b^2}{a^2}x^2 + x^2 - 2\,cx + c^2}$$

$$= \sqrt{c^2 + b^2 - 2\,cx + \frac{a^2 - b^2}{a^2}x^2}.$$

$$= \sqrt{a^2 - 2\,cx + \frac{c^2}{a^2}x^2}$$

$$= a - \frac{c}{a}x\,;$$

hence $r = a - ex.$ See (1) Art. 76 . . . (1)

Similarly we find

$$r' = a + ex \ . \ . \ . \ (2)$$

78. Having given the transverse axis and the foci of any ellipse, the principles of Art. 75 enables us to construct the ellipse by three different methods.

First Method. — Take a cord equal in length to the transverse axis AA'. Attach one end of it at F, the other at F'. Place the point of a pencil in the loop formed by the cord and stretch it upward until taut. Wheeling the pencil around, while keeping the point on the paper and tightly pressed against the cord, the path described will be an arc of the ellipse. After describing the upper half of the ellipse, remove the pencil and form the loop below the transverse axis. By a similar process the lower half may be described. It is

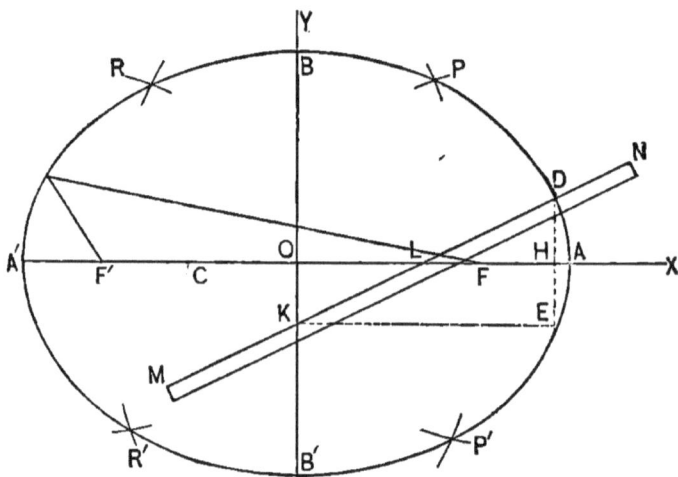

FIG. 36.

evident during the operation that the sum of the distances of the point of the pencil from the foci is constant and equal to the length of the cord; i.e., to the transverse axis.

Second Method. — Take any point C on the transverse axis and measure the distances A'C, AC. With F' as a centre and CA' as a radius describe the arc of a circle; also with F as a centre and CA as a radius describe another arc. The points R, R' in which these arcs intersect are points of the ellipse. By interchanging the radii two other points P, P' may be determined. A smooth curve traced through a number of points thus located will be the required ellipse.

Third Method. — Let the axes AA' = 2 a, BB' = 2 b be given. Lay off on any straight edge MN (a piece of paper will do) KD = OA = a and DL = OB = b. Place the straight edge on the axes in the position indicated in the figure. Then as K and L slide along the axes, the point D will describe the ellipse. For from the figure DLH and DKE are similar triangles:

$$\therefore \frac{DK}{KE} = \frac{DL}{LH}; \text{ i.e., } \frac{a}{x} = \frac{b}{\sqrt{b^2 - y^2}} \text{ (x and y being}$$

the co-ordinates of D).

Hence, squaring, clearing of fractions, and transposing, we have

$$a^2 y^2 + b^2 x^2 = a^2 b^2.$$

That is the locus described by D is an ellipse. An instrument based upon this principle is commonly used for drawing the ellipse.

79. *To find the latus rectum, or parameter of an ellipse.*

The latus rectum or parameter of an ellipse is the double ordinate passing through the focus.

The abscissas of the points in which the latus rectum pierces the ellipse are $x = \pm \sqrt{a^2 - b^2}$. Substituting *either* of these values on the equation of the ellipse

$$y^2 = \frac{b^2}{a^2} (a^2 - x^2),$$

we have $\qquad y^2 = \frac{b^2}{a^2} (a^2 - (a^2 - b^2)) = \frac{b^4}{a^2} \therefore y = \frac{b^2}{a}.$

Hence $\qquad$ Latus rectum $= 2y = \dfrac{2b^2}{a} \ldots$ (1)

Forming a proportion from this equation there results,

$$2y : 2b :: b : a;$$

hence $\qquad 2y : 2b :: 2b : 2a;$

i.e., *the latus rectum is a third proportional to the two axes.*

EXAMPLES.

Find the semi-axes, the eccentricity, and the latus rectum of each of the following ellipses:

1. $3x^2 + 2y^2 = 6.$ $\qquad$ 3. $x^2 + 3y^2 = 2.$

2. $\dfrac{x^2}{3} + \dfrac{y^2}{2} = 1.$ $\qquad$ 4. $4y^2 + 6 = 8 - 2x^2.$

5. $ax^2 + by^2 = ab.$ **7.** $y^2 + \dfrac{x^2}{2} = m.$

6. $cy^2 + x^2 = d.$ **8.** $x^2 + \dfrac{y^2}{m} = n.$

Write the equation of the ellipse having given:

9. The transverse axis $= 10$; the distance between the foci $= 8$.

$$Ans. \quad \frac{x^2}{25} + \frac{y^2}{9} = 1.$$

10. Sum of the axes $= 18$; difference of axes $= 6$.

$$Ans. \quad \frac{x^2}{36} + \frac{y^2}{9} = 1.$$

11. Transverse axis $= 10$; the conjugate axis $= \frac{1}{2}$ the transverse axis.

$$Ans. \quad \frac{x^2}{25} + \frac{4\,y^2}{25} = 1.$$

12. Transverse axis $= 20$; conjugate axis $=$ distance between foci.

$$Ans. \quad \frac{x^2}{2} + y^2 = 50.$$

13. Conjugate axis $= 10$; distance between foci $= 10$.

$$Ans. \quad \frac{x^2}{2} + y^2 = 25.$$

14. Given $3\,y^2 + 4\,x^2 = 12$; required the co-ordinates of the point whose ordinate is double its abscissa.

$$Ans. \quad \left(\sqrt{\frac{6}{8}}, \, 2\sqrt{\frac{6}{8}} \right).$$

15. Given the ellipse $3\,y^2 + 2\,x^2 = 12$, and the line $y = x - 1$; to find the co-ordinates of their points of intersection.

16. Given the ellipse $\dfrac{x^2}{64} + \dfrac{y^2}{15} = 1$, and the abscissa of a point on the curve $= \frac{1}{2}$; required the focal radii of the point.

$$Ans. \quad r = 7\tfrac{9}{16}, \ r' = 8\tfrac{7}{16}.$$

THE ELLIPSE. 113

80. *To deduce the polar equation of the ellipse, either focus being taken as the pole.*

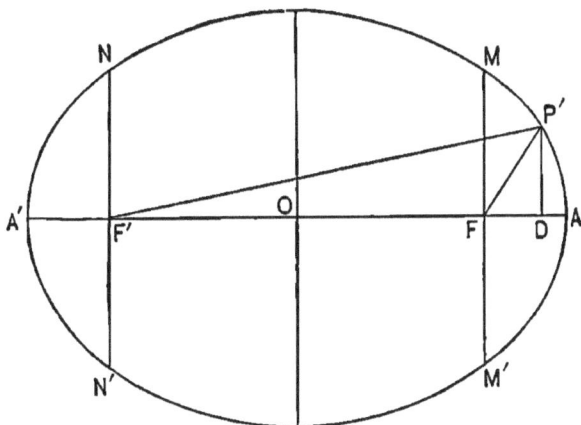

FIG. 37.

Let us take F as the pole, and let $(FP', P'FA) = (r, \theta)$ be the co-ordinates of any point P' of the ellipse. From Art. 77 (1) we have, $r = a - ex' \ldots$ (1)

From the figure, $OD = OF + FD$;

i.e., $\qquad x' = ae + r \cos \theta.$

Substituting this value of x' in (1), we have

$$r = a - e(ae + r \cos \theta),$$

or, reducing, we have

$$r = \frac{a(1 - e^2)}{1 + e \cos \theta} \ldots (2)$$

for the polar equation of the ellipse, the right-hand focus being taken as the pole.

From Art. 77 (2),

$$F'P' = r' = a + ex'.$$

We readily determine from this value

$$r' = \frac{a(1 - e^2)}{1 - e \cos \theta} \ldots (3)$$

for the polar equation of the ellipse, the left-hand focus being taken as the pole.

Cor. If $\qquad \theta = 0, \qquad r = a\,(1 - e) = \text{FA},$
$$r' = a\,(1 + e) = \text{F}'\text{A}.$$

If $\qquad \theta = 90°, \qquad r = a\,(1 - e^2) = a - a\,\dfrac{a^2 - b^2}{a^2}$

$$= \frac{b^2}{a} = \text{FM}.$$

$$r' = a\,(1 - e^2) = a - a\,\frac{a^2 - b^2}{a^2} = \frac{b^2}{a} = \text{F}'\text{N}.$$

If $\qquad \theta = 180°, \quad r = a\,(1 + e) = \text{FA}',$
$$r' = a\,(1 - e) = \text{F}'\text{A}'.$$

If $\qquad \theta = 270°, \quad r = a\,(1 - e^2) = \text{FM}',$
$$r' = a\,(1 - e^2) = \text{F}'\text{N}'.$$

If $\qquad \theta = 360°, \quad r = a\,(1 - e) = \text{FA},$
$$r' = a\,(1 + e) = \text{F}'\text{A}.$$

81. *To deduce the equation of condition for the supplemental chords of an ellipse.*

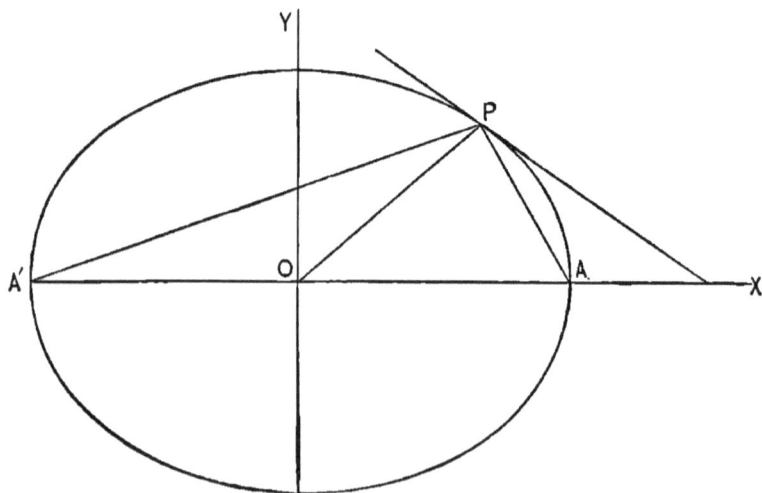

Fig. 38.

Let AP, A'P be a pair of supplemental chords.
The equation of a line through A (a, o) is

$$y = s\,(x - a).$$

The equation of a line through A' $(-a, o)$ is

$$y = s' (x + a).$$

Where these lines intersect we must have

$$y^2 = ss' (x^2 - a^2) \ldots (1)$$

In order that the lines shall intersect *on* the ellipse their equations must subsist at the same time with the equation of the ellipse

$$y^2 = \frac{b^2}{a^2} (a^2 - x^2) \ldots (2)$$

Dividing (1) by (2), we have

$$1 = - \frac{a^2}{b^2} ss';$$

or

$$ss' = - \frac{b^2}{a^2} \ldots (3)$$

for the required condition.

Cor. If $a = b$, the ellipse becomes a circle and (3) becomes

$$ss' = -1,$$

a relationship heretofore deduced. Art. 40 (1).

Schol. The preceding discussions have developed a remarkable analogy between the ellipse and circle. As we proceed we shall find that the circle is only a particular form of the ellipse and that all of the equations pertaining to it may be deduced directly from the corresponding equations deduced for the ellipse by simply making $a = b$ in those equations.

82. *To deduce the equation of the tangent to the ellipse.*

Let P'' (x'', y''), P' (x', y') be the points in which a secant P''S cuts the ellipse. Its equation is, therefore,

$$y - y' = \frac{y' - y''}{x' - x''} (x - x') \ldots (1)$$

As the points are on the ellipse, we must have

$$y'^2 = \frac{b^2}{a^2} (a^2 - x'^2) \ldots (2)$$

$$y''^2 = \frac{b^2}{a^2} (a^2 - x''^2) \ldots (3)$$

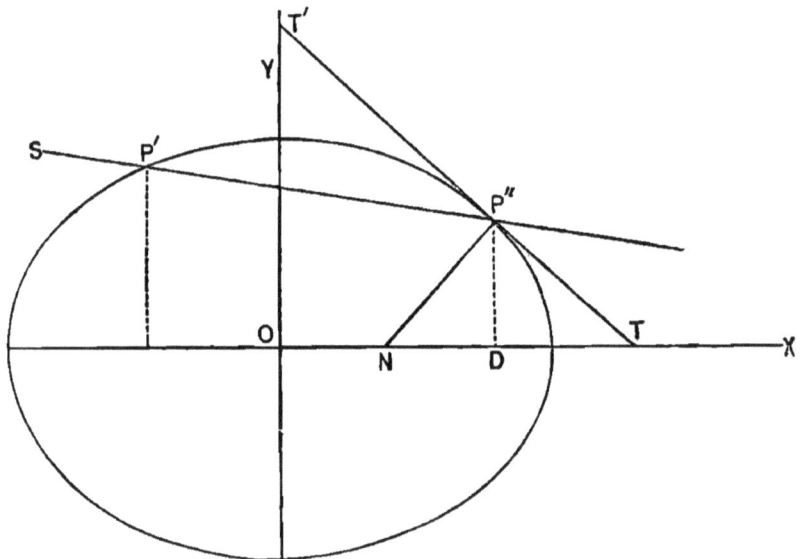

FIG. 39.

These three equations must subsist at the same time; hence subtracting (3) from (2) and factoring, we have

$$(y' - y'')(y' + y'') = -\frac{b^2}{a^2}(x' - x'')(x' + x'');$$

hence

$$\frac{y' - y''}{x' - x''} = -\frac{b^2}{a^2} \cdot \frac{x' + x''}{y' + y''}.$$

Substituting this value in (1) it becomes

$$y - y' = -\frac{b^2}{a^2} \cdot \frac{x' + x''}{y' + y''}(x - x').$$

Revolving the secant line upward about the point P'' (x'', y'') the other point of intersection P' (x', y') will approach P'' and will finally coincide with it. When this occurs the secant becomes a tangent and $x' = x''$, $y' = y''$; hence, substituting, we have

$$y - y'' = -\frac{b^2}{a^2} \cdot \frac{x''}{y''}(x - x'');$$

i.e.,

$$a^2 yy'' + b^2 xx'' = a^2 b^2; \quad \dots \quad (4)$$

or

$$\frac{xx''}{a^2} + \frac{yy''}{b^2} = 1 \quad \dots \quad (5)$$

for *the equation of the tangent.*

Cor. If $b = a$, we have

$$\frac{xx''}{a^2} + \frac{yy''}{a^2} = 1$$

for the equation of the tangent to the circle. See Art. 41 (6).

Schol. If we make x and y successively $= 0$ in the equation of the tangent (5), we have $y = \frac{b^2}{y''}$ and $x = \frac{a^2}{x''}$ for the values of the variable intercepts OT', OT, Fig. 39;

hence $\quad y'' = \frac{b^2}{y}$ and $x'' = \frac{a^2}{x}$.

These values in the equation

$$\frac{x''^2}{a^2} + \frac{y''^2}{b^2} = 1$$

give, after reduction,

$$\frac{a^2}{x^2} + \frac{b^2}{y^2} = 1 \ldots (6)$$

for the equation of the ellipse, the intercepts of its tangents on the axes being the variables.

83. *To deduce the value of the sub-tangent.*

Making $y = 0$ in (5), Art. 82, we have

$$x = OT = \frac{a^2}{x''};$$

$\therefore$ sub-tangent $= DT = \frac{a^2}{x''} - x'' = \frac{a^2 - x''^2}{x''}$.

Cor. If $b = a$, then from Art. 41, Schol. $a^2 - x''^2 = y''^2$,

$\therefore$ sub-tangent in the circle $= \frac{y''^2}{x''}$.

Schol. The value of the sub-tangent being independent of the value of the minor axis (2 b) it follows that this value is the same for every ellipse which is concentric with the given ellipse, and whose common transverse axis is 2 a.

84. The equation of condition that a line shall pass through the centre of the ellipse and the point of tangency is, Fig. 39,

$$y'' = tx'',$$

∴ the slope of this line is

$$t = \frac{y''}{x''}.$$

The slope of the tangent at (x'', y'') is, Art. 82,

$$t' = -\frac{b^2}{a^2} \cdot \frac{x''}{y''}.$$

Multiplying, member by member, we have

$$tt' = -\frac{b^2}{a^2} \dots (1)$$

But Art. 81 (3)

$$ss_{,} = -\frac{b^2}{a^2}$$

$$\therefore ss' = tt';$$

i.e., *the tangent to the ellipse and the line joining the centre and the point of tangency enjoy the property of being supplemental chords of an ellipse whose semi-axes bear to each other the ratio $\frac{b}{a}$.*

Cor. If $s = t$, then $s' = t'$; i.e., *if one supplementary chord is parallel to a diameter of the ellipse, the other supplementary chord is parallel to the tangent drawn at the extremity of that diameter.*

85. The principles of Arts. 83, 84 afford us two different methods of constructing a tangent to the ellipse at a given point.

First Method. — Art. 83, Schol.   Let P″, Fig. 40, be the given point.   Through P″ draw the ordinate P″D and produce it until it meets the circle described upon the transverse axis of the ellipse (AA′) in P′; draw P′T tangent to the circle at P′.   Join P″ and T; P″T will be the required tangent.

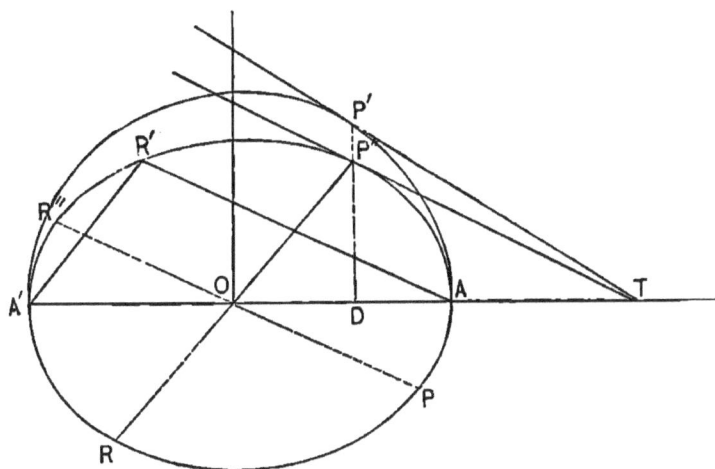

FIG. 40.

Second Method. — Art. 84 and Cor. Draw P″R through the centre, and from A′ draw A′R′ ∥ to P″R ; P″′T drawn through P″ ∥ to R′A will be tangent to the ellipse at P″.

86. *To deduce the equation of the normal to the ellipse.*
The equation of any line through P″ ($x″$, $y″$), Fig. 39, is

$$y - y″ = s (x - x″) \ . \ . \ . \ (1).$$

In order that this line and the tangent at P″ ($x″$, $y″$) shall be perpendicular their slopes must satisfy the condition

$$1 + ss′ = 0 \ . \ . \ . \ (2).$$

We have found Art. 82 for the slope of the tangent

$$s′ = - \frac{b^2}{a^2} \cdot \frac{x″}{y″} ;$$

hence, the slope of the normal is

$$s = \frac{a^2}{b^2} \cdot \frac{y″}{x″} .$$

Substituting this value of s in (1), we have

$$y - y″ = \frac{a^2 y″}{b^2 x″} (x - x″) \ . \ . \ . \ (3).$$

for the *equation of the normal to the ellipse.*

Cor. 1. If $a = b$, then (3) becomes, after reduction,
$$yx'' - xy'' = 0,$$
which is the equation of the normal line to the circle.

87. *To deduce the value of the sub-normal.*

Making $y = 0$ in the equation of the normal, (3), Art. 86,

we have, Fig. 39, $ON = x = \dfrac{a^2 - b^2}{a^2} x'' = e^2 x''$,

∴ Sub-normal $= DN = x'' - \dfrac{a^2 - b^2}{a^2} x'' = \dfrac{b^2}{a^2} x''.$

Cor. 1. If $a = b$, then
$$\text{Sub-normal for the circle} = x''.$$

EXAMPLES.

1. Deduce the polar equation of the ellipse, the pole being at the centre and the initial line coincident with the X-axis.

$$Ans. \quad r = \frac{ab}{\sqrt{a^2 \sin^2 \theta + b^2 \cos^2 \theta}}.$$

Write the equation of the tangent to each of the following ellipses, and give the value of the sub-tangent in each case.

2. $2x^2 + 4y^2 = 38$ at $(1, 3)$.

$$Ans. \quad x + 6y = 19; \; 18.$$

3. $\dfrac{x^2}{3} + \dfrac{y^2}{2} = 1$, at (1, ordinate positive).

$$Ans. \quad x + \frac{3}{\sqrt{3}} y = 3; \; 2.$$

4. $\dfrac{x^2}{4} + \dfrac{y^2}{3} = 1$, at $(2, 0)$.

$$Ans. \quad x = 2; \; 0.$$

5. $2x^2 + 3y^2 = 11$ at $(2, -1)$.

$$Ans. \quad 4x - 3y = 11; \; \tfrac{3}{4}.$$

6. $\dfrac{y^2}{a} + \dfrac{x^2}{b} = 1$, at $(0, \sqrt{a})$.

7. $\dfrac{x^2}{a^2} + \dfrac{y^2}{b^2} = 1,$ at (a, o)

8. $y^2 + bx^2 = 2,$ at $(1, -\sqrt{2-b})$.

9. $\dfrac{x^2}{m} + y^2 = 1,$ at $(abs\ +, .5)$.

Write the equation of the normal to each of the following ellipses, and give the value of the sub-normal.

10. $3y^2 + 4x^2 = 39,$ at $(3, 1)$.

11. $4y^2 + 2x^2 = 44,$ at $(-2, \text{ord negative})$.

12. $\dfrac{x^2}{4} + \dfrac{y^2}{2} = 1,$ at $(-1, \text{ord} -)$.

13. $\dfrac{x^2}{3} + \dfrac{y^2}{6} = 1,$ at $(1, 2)$.

14. $\dfrac{x^2}{a} + y^2 = 1,$ at $(\tfrac{1}{2}, \text{ord} +)$.

15. $m^2 y^2 + n^2 x^2 = m^2 n^2,$ at (m, o).

16. The equation of a chord of an ellipse is $y = -2x + 6$; what is the equation of the supplementary chord, the axes of the ellipse being 6 and 4 ?

Ans. $y = \tfrac{2}{9}x + \tfrac{6}{5}$.

17. Given the equation $\dfrac{x^2}{9} + \dfrac{y^2}{16} = 1,$ and $y - 2 = 0$; required the equation of the tangents to the ellipse at the points in which the line cuts the curve.

18. Given the ellipse $\dfrac{x^2}{4} + \dfrac{y^2}{9} = 1,$ and the line $y - x + 2 = 0$; required

(*a*) The equation of a tangent to the ellipse ∥ to the line.

(*b*) "   "   "   "   ,   "   "   ⊥ "   "   "

19. The point (4, 3) is outside the ellipse

$$\frac{x^2}{16} + \frac{y^2}{9} = 1\,;$$

required the equations of the tangents to the ellipse which pass through the point.

88. *The angle formed by the focal lines drawn to any point of an ellipse is bisected by the normal at that point.*

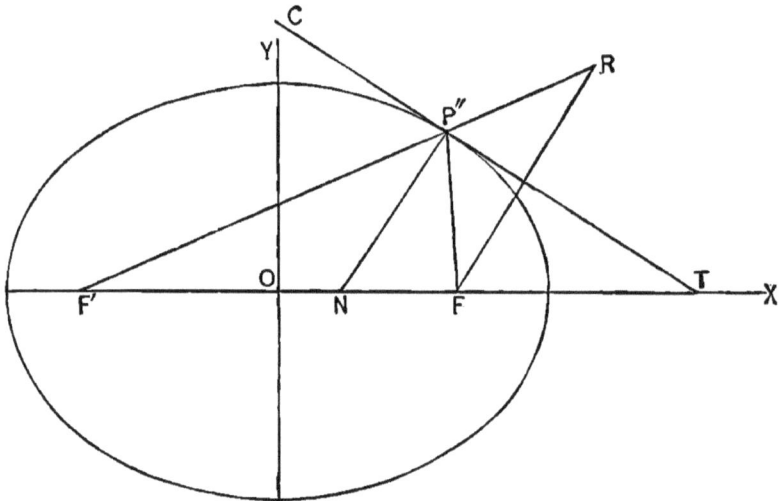

FIG. 41.

Let P″N be a normal at any point P″ (x'', y''). Draw P″F, P″F.′

We have found, Art. 87, that

$$ON = \frac{a^2 - b^2}{a^2}\, x'' = e^2 x''.$$

From Art. 76 we have OF = OF′ = ae; hence

$$NF = OF - ON = ae - e^2 x'' = e\,(a - ex'')$$
$$NF' = OF' + ON = ae + e^2 x'' = e\,(a + ex'')$$
$$\therefore\ NF : NF' :: (a - ex'') : (a + ex'')$$

But FP″ : F′P″ :: $(a - ex'') : (a + ex'')$ Art. 77, (1) and (2);

$$\therefore\ NF : NF' :: FP'' : F'P''.$$

The normal, therefore, divides the base of the triangle F'P''F into two segments which are proportional to the adjacent sides. Hence

$$FP''N = F'P''N.$$

SCHOL. 1. If P'''T be a tangent drawn at P''', we must have

$$F'P''C = FP'''T;$$

for each of these angles is equal to the difference between a right angle and the angle F'P''N (= FP''N). Hence, the *tangent to the ellipse makes equal angles with the focal radii drawn to the point of tangency.*

SCHOL. 2. The principles of this article afford us another method of drawing a tangent to the ellipse at a given point. Let P'' be a point at which we wish to draw a tangent. Produce F'P'' to R, making P''R = FP''; join F and R. A line P''T, drawn through P'' $\perp$ to FR will be tangent to the ellipse at P''.

89. *To find the condition that the straight line* $y = sx + c$ *must fulfil in order that it may touch the ellipse*

$$\frac{x^2}{a^2} + \frac{y^2}{b^2} = 1.$$

If we consider the line as a secant and combine the equations

$$y = sx + c,$$
$$\frac{x^2}{a^2} + \frac{y^2}{b^2} = 1,$$

we obtain the co-ordinates of the points of intersection. Eliminating y from these equations, we have

$$x = \frac{-sa^2c \pm ab\sqrt{s^2a^2 + b^2 - c^2}}{s^2a^2 + b^2} \quad \ldots \text{(1)}$$

for the abscissas of the points of intersection. Now, when the secant line becomes a tangent, these abscissas become equal. Looking at (1) we see that the condition for equality of abscissas is that the radical in the numerator shall disappear; hence

$$s^2a^2 + b^2 - c^2 = 0,$$

or $\qquad s^2a^2 + b^2 = c^2 \ \ldots \ (2)$

is the required condition.

Cor. If we substitute the value of c drawn from (2) in the equation of the line, we have

$$y = sx \pm \sqrt{s^2a^2 + b^2} \ \ldots \ (3)$$

for the equation of the tangent to the ellipse in terms of its slope.

90. *To find the locus generated by the intersection of a tangent to the ellipse and a perpendicular to it from a focus as the point of tangency moves around the curve.*

The equation of a straight line through the focus (ae, o) is

$$y = s' (x - ae).$$

In order that this line shall be perpendicular to the tangent

$$y = sx \pm \sqrt{s^2a^2 + b^2} \ \ldots \ (1),$$

its equation must be

$$y = -\frac{1}{s} (x - ae) \ \ldots \ (2)$$

If we now combine (1) and (2) so as to eliminate the slope (s), the resulting equation will express the relationship between the co-ordinates of the point of intersection of these lines in every position they may assume; hence it will be the equation of the required locus.

Transposing sx to the first member in (1), and clearing (2) of fractions and transposing, we have

$$y - sx = \pm \sqrt{s^2a^2 + b^2}.$$
$$sy + x = ae.$$

Squaring these equations and adding, remembering that $a^2 - b^2 = a^2e^2$, Art. 76, we have,

$$(1 + s^2)(x^2 + y^2) = (1 + s^2) a^2,$$

or $\qquad x^2 + y^2 = a^2; \ \ldots \ (3);$

hence, *the circle constructed on the transverse axis of the ellipse is the locus of the intersection of the tangents and the perpendiculars let fall from the focus on them.*

This circle is known as the Major-Director circle of the ellipse. (See Fig. 45.)

91. *To find the locus generated by the intersection of two tangents which are perpendicular to each other as the points of tangency more around the curve.*

The equation of a tangent to the ellipse is

$$y = sx + \sqrt{s^2a^2 + b^2} \ldots (1)$$

The equation of a tangent perpendicular to (1) is

$$y = -\frac{1}{s}x + \sqrt{\frac{a^2}{s^2} + b^2}; \ldots (2)$$

hence, by a course of reasoning analogous to that of the preceding article, we have

$$x^2 + y^2 = a^2 + b^2 \ldots (3)$$

The required locus is, therefore, *a circle concentric with the ellipse and having its radius equal to* $\sqrt{a^2 + b^2}$.

92. *Two tangents are drawn to the ellipse from a point without; required the equation of the line joining the points of tangency.*

Let P' (x', y'), Fig. 42, be the given point, and let P'' (x'', y''), P_2 (x_2, y_2) be the points of tangency. Since P' (x', y') is a point common to both tangents, its co-ordinates must satisfy their equations; hence,

$$\frac{x'x''}{a^2} + \frac{y'y''}{b^2} = 1.$$

$$\frac{x'x_2}{a^2} + \frac{y'y_2}{b^2} = 1.$$

Hence (x'', y'') and (x_2, y_2) will satisfy the equation

$$\frac{x'x}{a^2} + \frac{y'y}{b^2} = 1 \ldots (1)$$

As (1) is the equation of a straight line, and is satisfied for the co-ordinates of both points of tangency, it must be the equation of the straight line which joins them.

93. *To find the equation of the polar of the pole* (x', y'), *with regard to the ellipse*

$$\frac{x^2}{a^2} + \frac{y^2}{b^2} = 1.$$

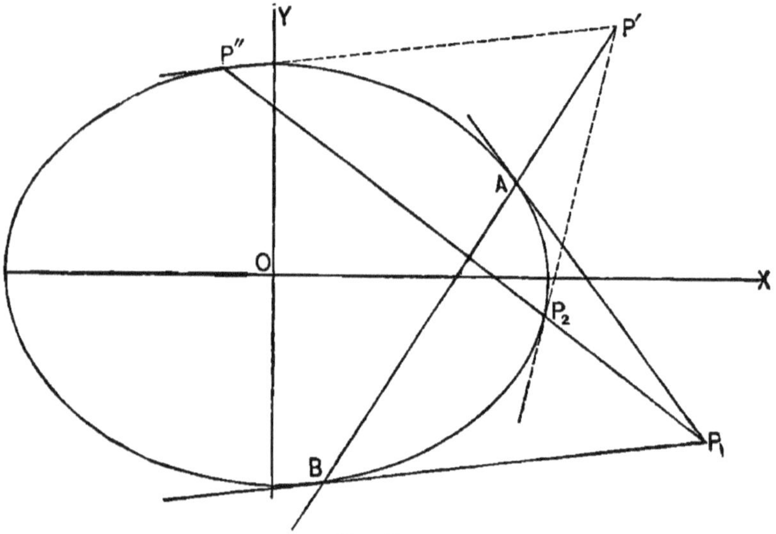

Fig. 42.

By the aid of Fig. 42, and a course of reasoning similar to that of Art. 49, the equation of P_1P'', the polar to P', may be shown to be

$$\frac{x'x}{a^2} + \frac{y'y}{b^2} = 1.$$

Cor. If the polar of the point P' (x', y') passes through P_1 (x_1, y_1), then the polar of P_1 (x_1, y_1) will pass through P' (x', y'). (See Art. 50.)

94. *To deduce the equation of the ellipse when referred to a pair of conjugate diameters as axes.*

A pair of conjugate diameters of the ellipse are those diam-

eters to which if the ellipse be referred its equation will contain only the second powers of the variables.

The equation of the ellipse when referred to its centre and axes is

$$\frac{x^2}{a^2} + \frac{y^2}{b^2} = 1 \ \ldots \ (1)$$

If we refer the ellipse to a pair of oblique axes having the origin at the centre, we have, Art. 33, Cor. 1,

$$x = x' \cos \theta + y' \cos \varphi$$
$$y = x' \sin \theta + y' \sin \varphi$$

for the equations of transformation. Substituting in (1), we have

$$(a^2 \sin^2 \theta + b^2 \cos^2 \theta)\, x'^2 + (a^2 \sin^2 \varphi + b^2 \cos^2 \varphi)\, y'^2$$
$$+ 2\,(a^2 \sin \theta \sin \varphi + b^2 \cos \theta \cos \varphi)\, x'y' = a^2 b^2 \ \ldots \ (2)$$

for the equation of the ellipse referred to oblique axes. But, by definition, the equation of the ellipse when referred to a pair of conjugate diameters contains *only* the second powers of the variables; hence

$$a^2 \sin \theta \sin \varphi + b^2 \cos \theta \cos \varphi = 0 \ \ldots \ (3)$$

is the condition that a pair of axes must fulfil in order to be conjugate diameters of the ellipse.

Making the co-efficient of $x'y'$ equal to zero in (2), we have after dropping accents

$$(a^2 \sin^2 \theta + b^2 \cos^2 \theta)\, x^2 + (a^2 \sin^2 \varphi + b^2 \cos^2 \varphi)\, y^2 = a^2 b^2 \ \ldots \ (4)$$

for the equation of the ellipse when referred to a pair of conjugate diameters. This equation, however, takes a simpler form when we introduce the semi-conjugate diameters. Making $y = 0$ and $x = 0$, successively, in (4), we have

$$\left.\begin{aligned} x^2 &= \frac{a^2 b^2}{a^2 \sin^2 \theta + b^2 \cos^2 \theta} = a'^2 \\ y^2 &= \frac{a^2 b^2}{a^2 \sin^2 \varphi + b^2 \cos^2 \varphi} = b'^2 \end{aligned}\right\} \ \ldots \ (5)$$

in which a' and b' represent the semi-conjugate axes. From (5), we have

$$a^2 \sin^2 \theta + b^2 \cos^2 \theta = \frac{a^2 b^2}{a'^2} \, ;$$

$$a^2 \sin^2 \varphi + b^2 \cos^2 \varphi = \frac{a^2 b^2}{b'^2} \, .$$

Substituting these values of the co-efficients in (4), we have, after reduction,

$$\frac{x^2}{a'^2} + \frac{y^2}{b'^2} = 1 \, . \, . \, . \, (6)$$

for the required equation.

Cor. As equation (6) contains only the second powers of the variables, it follows that each of the two diameters to which the curve is referred will bisect all chords drawn parallel to the other.

Schol. The equation of condition for conjugate diameters (3) may be put under the forms

$$\tan \theta \tan \varphi = -\frac{b^2}{a^2} \, . \, . \, . \, (7)$$

Comparing this expression with (3) Art. 81, we see that the same result was obtained for the supplementary chords of an ellipse; hence, Fig. 40, if A'R', R'A be a pair of supplementary chords, then RP'', PR'', drawn through the centre parallel to these chords, will be a pair of conjugate diameters. Again: comparing (7) with (1) Art. 84, we see that the same relationship was obtained for a diameter and the tangent drawn at its extremity; hence, Fig. 40, if P''R be a diameter and P'''T be a tangent drawn at its extremity, then PR'', drawn through the centre parallel to P'''T, is the conjugate diameter to RP''.

The equation of condition (7) being a single equation containing two unknown quantities (tan θ, tan φ), we may assume any value we please for one of them, and the equation will make known the value of the other; hence, *in the ellipse there are an infinite number of pairs of conjugate diameters.*

95. *To find the equation of a conjugate diameter.*

Let P"R, R'P' be a pair of conjugate diameters. We wish to find the equation of R'P'.

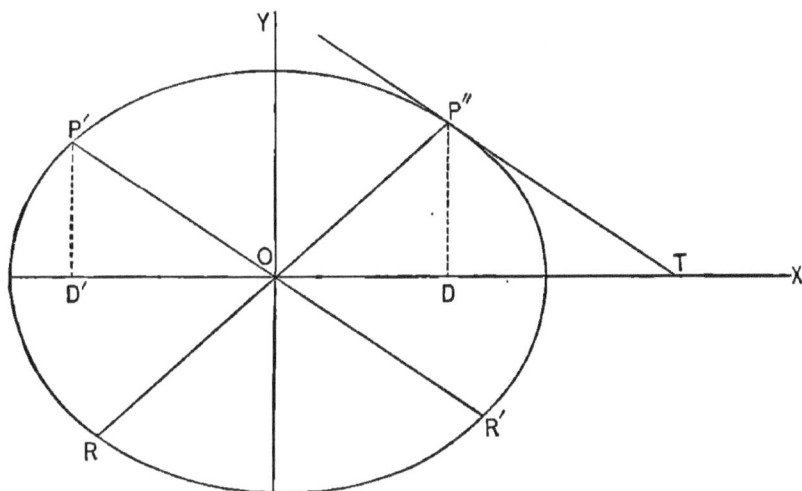

FIG. 43.

The equation of the tangent line P"T, drawn through P" (x'', y'') is

$$\frac{xx''}{a^2} + \frac{yy''}{b^2} = 1.$$

By Art. 94, Schol., the diameter P'R' is parallel to P"T; hence its equation must be the same as that of the tangent, the constant term being zero.

$$\therefore \frac{xx''}{a^2} + \frac{yy''}{b^2} = 0 \ \ldots \ (1)$$

or

$$y = -\frac{b^2x''}{a^2y''} x \ \ldots \ (2)$$

is the equation of a diameter expressed in terms of the co-ordinates of the extremity of its conjugate diameter.

Cor. Let s represent the slope of the diameter P"R; then, from (2)

$$y = -\frac{b^2x''}{a^2y''} x = -\frac{b^2}{a^2} \cdot \frac{1}{s} \, x,$$

since
$$\frac{OD}{DP''} = \frac{x''}{y''} = \frac{1}{s};$$

hence we have

$$y = -\frac{b^2}{a^2 s} x \; \ldots \; (3)$$

for the equation of a diameter in terms of the slope of its conjugate diameter.

96. *To find the co-ordinates of either extremity of a diameter, the co-ordinates of one extremity of its conjugate diameter being given.*

Let P''R and R'P', Fig. 43, be a pair of conjugate diameters. Let (x'', y'') be the co-ordinates of P''. We wish to find the co-ordinates (x', y') of P' in terms of the co-ordinates of P.''

The equation of condition that P' (x', y') shall be on the diameter P'R' is, Art. 95, (1)

$$\frac{x'x''}{a^2} + \frac{y'y''}{b^2} = 0.$$

Since P' (x', y') is on the ellipse, we have also

$$\frac{x'^2}{a^2} + \frac{y'^2}{b^2} = 1.$$

Eliminating y' and x', successively, from these equations, we find

$$x' = \mp \frac{a}{b} y'' \text{ and } y' = \pm \frac{b}{a} x''.$$

These expressions, taken with the upper signs, are the co-ordinates of P'; taken with the lower signs, they are the co-ordinates of R'.

97. *To show that the sum of the squares on any pair of semi-conjugate diameters is equivalent to the sum of the squares on the semi-axes.*

Let P'' (x'', y'') and P' (x', y'), Fig. 43, be the extremities of any two semi-conjugate diameters. Let $OP'' = a'$, $OP' = b'$; then, from the triangles ODP'', $OD'P'$, we have,

$$a'^2 = x''^2 + y''^2 \cdots (1)$$

and $\qquad b'^2 = x'^2 + y'^2 \cdots (2)$

But, Art. 96, $\quad x'^2 = \dfrac{a^2}{b^2} y''^2,$

and $\qquad y'^2 = \dfrac{b^2}{a^2} x''^2;$

hence $\qquad b'^2 = \dfrac{a^2}{b^2} y''^2 + \dfrac{b^2}{a^2} x''^2 \cdots (3).$

Adding (1) and (3), we have

$$a'^2 + b'^2 = (a^2 + b^2)\left(\dfrac{x''^2}{a^2} + \dfrac{y''^2}{b^2}\right);$$

but $\qquad \dfrac{x''^2}{a^2} + \dfrac{y''^2}{b^2} = 1;$

hence, $\qquad a'^2 + b'^2 = a^2 + b^2 \cdots (4)$

98. *To show that the parallelogram constructed on any two conjugate diameters is equivalent to the rectangle constructed on the axes.*

Let $P''R \; (= 2a')$, $P'R' \; (= 2b')$, Fig. 44, be any two conjugate diameters. To prove that area $CTC'T' =$ area $BB'H'H$.

The area of the parallelogram $OP'''TR'$ is

$$OR' \times P''P.$$

From the figure $P''P = OP'' \sin P''OR'$
$= a' \sin (180° - (\varphi - \theta)) = a' \sin (\varphi - \theta);$
$\therefore$ area of $OP'''TR' = a'b' \sin (\varphi - \theta) \cdots (1)$

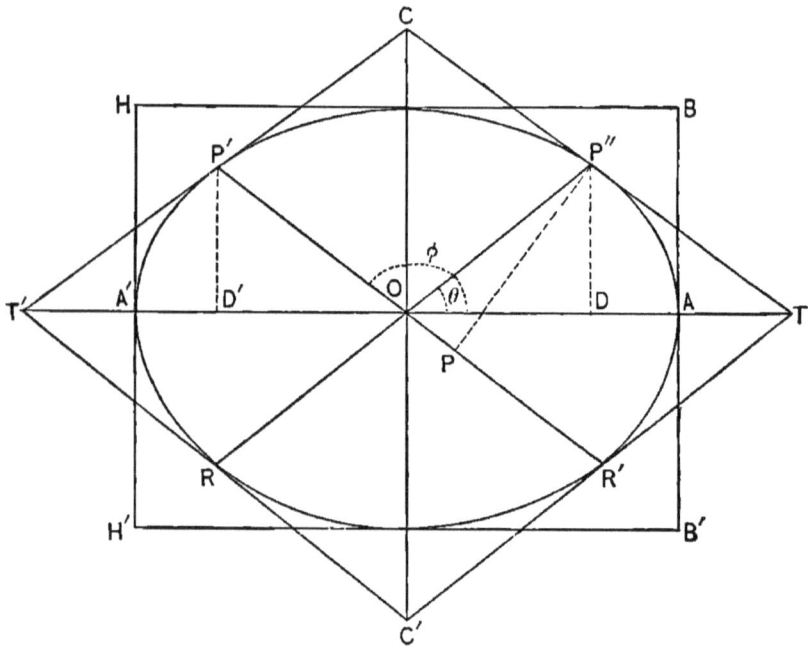

Fig. 44.

From the triangles OD'P', ODP'', we have

$$\sin \varphi = \frac{D'P'}{OP'} = \frac{y'}{b'} = \frac{bx''}{ab'} \; ; \; \sin \theta = \frac{y''}{a'}$$

$$\cos \varphi = \frac{OD'}{OP'} = -\frac{x'}{b'} = -\frac{ay''}{bb'} \; ; \; \cos \theta = \frac{x''}{a'} \cdot$$

Hence

$$\sin (\varphi - \theta) = \sin \varphi \cos \theta - \cos \varphi \sin \theta$$

$$= \frac{bx''^2}{aa'b'} + \frac{ay''^2}{ba'b'}$$

$$= \frac{b^2x''^2 + a^2y''^2}{aa'b'b} \cdot$$

$$= \frac{a^2b^2}{aba'b'} \cdot$$

$$= \frac{ab}{a'b'}$$

Substituting this value in (1) and multiplying through by 4, we have

$$\text{area OP}'''\text{TR}' \times 4 = 4\,ab;$$

i.e., $\qquad$ area CTC'T' = area BB'H'H.

99. *To show that the ordinate of any point on the ellipse is to the ordinate of the corresponding point on the circumscribing circle as the semi-conjugate axis of the ellipse is to the semi-transverse axis.*

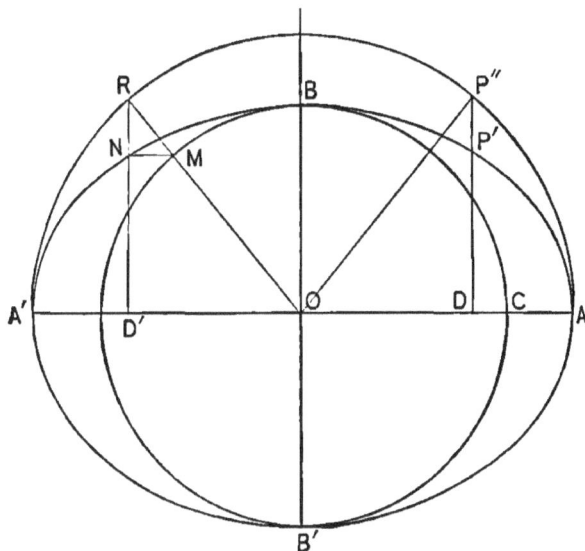

FIG. 45.

Let DP', DP'' be the ordinates of the corresponding points P' (x', y') and P'' (x'', y'').

Since P' (x', y') is on the ellipse, we have

$$y'^2 = \frac{b^2}{a^2}(a^2 - x'^2).$$

Since P'' (x'', y'') is on the circle whose radius is a, we have

$$y''^2 = a^2 - x''^2.$$

Dividing these equations, member by member, we have

$$\frac{y'^2}{y''^2} = \frac{b^2}{a^2}, \quad (\text{since } x' = x'');$$

$$\therefore y' : y'' :: b : a.$$

Similarly we may prove that
$$x_1 : x_2 :: a : b,$$
where x_1 is the abscissa of any point on the ellipse, and x_2 is the corresponding abscissa of a point on the *inscribed* circle.

100. The principles of the preceding article give us a method of describing the ellipse by points when the axes are given.

From O, Fig. 45, as a centre with radii equal to the semi-axes OA, OB describe the circles A'RA, BCB'. Draw any radius OR of the larger circle, cutting the smaller circle in M; draw MN ‖ to OA', cutting the ordinate let fall from R in N; N is a point of the ellipse. Since MN is parallel to the base of the triangle RD'O, we have

$$D'N : D'R :: OM : OR;$$
i. e.,
$$y' : y'' :: b : a;$$

hence, the construction.

101. *To show that the area of the ellipse is to the area of the circumscribing circle as the semi-minor axis of the ellipse is to its semi-major axis.*

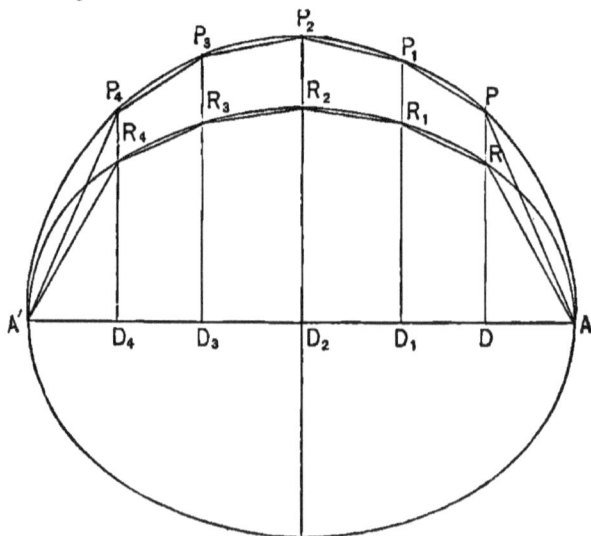

FIG. 46.

Inscribe in the ellipse any polygon $ARR_1R_2R_3R_4A'$, and from its vertices draw the ordinates RD, R_1D_1, etc., producing them upward to meet the circle in P, P_1, P_2, etc. Joining these points we form the inscribed polygon $APP_1P_2P_3P_4A'$ in the circle.

Let $(x, y_0), (x', y_1), (x'', y_2)$ etc., be the co-ordinates of P, P_1, P_2, etc., and let $(x, y), (x', y'), (x'', y'')$, etc., be the co-ordinates of the corresponding points R, R_1, R_2, etc., of the ellipse.

Then

$$\text{Area } RDD_1R_1 = (x - x')\frac{y + y'}{2}.$$

$$\text{Area } PDD_1P_1 = (x - x')\frac{y_0 + y_1}{2};$$

hence

$$\frac{\text{Area } RDD_1R_1}{\text{Area } PDD_1P_1} = \frac{y + y'}{y_0 + y_1}.$$

But, Art. 99, $\dfrac{y}{y_0} = \dfrac{b}{a}$ and $\dfrac{y'}{y_1} = \dfrac{b}{a}$;

$$\therefore \frac{y + y'}{y_0 + y_1} = \frac{b}{a}.$$

Hence

$$\frac{\text{Area } RDD_1R_1}{\text{Area } PDD_1P_1} = \frac{b}{a}.$$

We may prove in like manner that every corresponding pair of trapezoids bear to each other this constant ratio; hence, by the Theory of Proportion, the sum of all the trapezoids in the ellipse will bear to the sum of all the trapezoids in the circle the same ratio. Representing these sums by Σt and ΣT, respectively, we have

$$\frac{\Sigma t}{\Sigma T} = \frac{b}{a}.$$

As this relationship holds true for any number of trapezoids, it holds true for the *limits* to which the sum of the trapezoids of the ellipse and the sum of the trapezoids of the circle approach as the number of trapezoids increase.

But these limits are the area of the ellipse and the area of the circle; hence

$$\frac{area\ of\ ellipse}{area\ of\ circle} = \frac{b}{a}.$$

Cor. Since the area of the circle is πa^2, we have

$$\frac{area\ of\ ellipse}{\pi a^2} = \frac{b}{a}.$$

$$\therefore area\ of\ ellipse = \pi ab.$$

Since
$$\pi a^2 : \pi ab :: \pi ab : \pi b^2,$$

we see that *the area of the ellipse is a mean proportional between the areas of the circumscribed and inscribed circles.*

EXAMPLES.

1. What must be the value of c in order that the line $y = 2x + c$ may touch the ellipse

$$\frac{x^2}{4} + \frac{y^2}{9} = 1\ ?$$

Ans. $c = 5$.

2. The semi-transverse of an ellipse is 10; what must be the value of the semi-conjugate axis in order that the ellipse may touch the line $2y + x - 14 = 0$?

Ans. $b = \sqrt{24}$.

3. What are the equations of the tangents to the ellipse

$$\frac{x^2}{5} + \frac{y^2}{6} = 1,$$

whose inclination to X-axis $= 45°$?

4. The locus of the intersection of the tangents to the ellipse

$$\frac{x^2}{a^2} + \frac{y^2}{b^2} = 1$$

drawn at the extremities of conjugate diameters is an ellipse; required its equation.

Ans. $\dfrac{x^2}{a^2} + \dfrac{y^2}{b^2} = 2$.

5. Tangents are drawn from the point $(0, 8)$ to the ellipse

$$\frac{x^2}{4} + y^2 = 1 ;$$

required the equation of the line joining the points of tangency. *Ans.* $8\,y - 1 = 0.$

Required the polar of the point $(5, 6)$ with respect to the following ellipses:

6. $x^2 + 3\,y^2 = 9.$ **7.** $\dfrac{x^2}{5} + \dfrac{y^2}{8} = 1.$

$$\textbf{8.} \quad \frac{x^2}{a^2} + \frac{y^2}{b^2} = 1.$$

9. What are the polars of the foci?

$$\textit{Ans.} \quad x = \pm \frac{a}{e}.$$

10. What is the pole of $y = 3\,x + 1$ with respect to

$$\frac{x^2}{4} + \frac{y^2}{9} = 1?$$

Ans. $(-12, 9).$

11. The line $3\,y = 5\,x$ is a diameter of

$$\frac{x^2}{4} + \frac{y^2}{9} = 1 ;$$

required the equation of the conjugate diameter.
Ans. $20\,y + 27\,x = 0.$

12. A pair of conjugate diameters in the ellipse

$$\frac{x^2}{16} + \frac{y^2}{9} = 1$$

make angles whose tangents are $\dfrac{3}{4}$ and $-\dfrac{3}{4}$, respectively, with the X-axis; required their lengths.

13. What is the area of the ellipse

$$\frac{x^2}{4} + \frac{y^2}{10} = 1?$$

Ans. $2\,\pi\,\sqrt{10}.$

14. The minor axis of an ellipse is 10, and its area is equal to the area of a circle whose diameter is 16; what is the length of the major axis ?   *Ans.* 25⅗.

15. The minor axis of an ellipse is 6, and the sum of the focal radii to a point on the curve is 16; required the major axis, the distance between the foci, and the area.

GENERAL EXAMPLES.

1. What is the equation of the ellipse which passes through (2, 4) (— 2, 4), the centre being at the origin ?

2. The major axis of an ellipse is $= 18$, and the point (6, 4) is on the curve; required the equation of the ellipse.

3. The lines $y = -\frac{1}{2}x + 6$ and $y = \frac{1}{8}x + \frac{3}{2}$ are supplemental chords drawn from the extremities of the transverse axis of an ellipse; required the equation of the ellipse.

4. The minor axis of an ellipse is $= 12$, and the foci and centre divide the major axis into four equal parts; required the equation of the ellipse.

5. Assuming the equation of the ellipse show that the sum of the distances of any point on the ellipse from the foci is constant and $=$ to the transverse axis.

6. The sub-tangent for a point whose abscissa is 2 is $= 6$ in an ellipse whose eccentricity is $\frac{1}{4}$; required the equation of the ellipse.   *Ans.* $\frac{x^2}{16} + \frac{y^2}{15} = 1.$

7. What are the equations of the tangents to

$$\frac{x^2}{9} + \frac{y^2}{25} = 1$$

which form with the X-axis an equilateral triangle ?

8. Show that the tangents drawn at the extremities of any chord intersect on the diameter which bisects that chord.

9. What are the equations of the tangents drawn at the extremities of the latus-rectum ?

10. Show that the pair of diameters drawn parallel to the chords joining the extremities of the axes are equal and conjugate.

11. A chord of the ellipse

$$\frac{x^2}{16} + \frac{y^2}{9} = 1$$

passes through the point (2, 3) and is bisected by the line $y - x = 0$; required the equation of the chord.

12. What are the equations of the pair of conjugate diameters of the ellipse $16\, y^2 + 9\, x^2 = 144$ which are equal ?

13. Show that either focus of an ellipse divides the major axis in two segments whose rectangle is equal (*a*) to the rectangle of the semi-major axis and semi-parameter; (*b*) to the square of the semi-minor axis.

14. Show that the rectangle of the perpendiculars let fall from the foci on a tangent is constant and equal to the square of the semi-minor axis.

15. A system of parallel chords which make an angle whose tangent $= 2$ with the X-axis are bisected by the diameter of an ellipse whose semi-axes are 4 and 3; required the equation of the diameter.

16. Show that the polar of a point on any diameter is parallel to the conjugate diameter.

17. Find the locus of the vertex of a triangle having given the base $= 2\,a$, and the product of the tangent of the angles at the base $= \dfrac{b^2}{c^2}$.

Ans. $b^2x^2 + c^2y^2 = b^2a^2$.

18. Find the locus of the vertex of a triangle having given the base $= 2\,a$, and the sum of the sides $= 2\,b$.

$$\text{Ans.} \quad \frac{x^2}{b^2} + \frac{y^2}{b^2 - a^2} = 1.$$

19. Find the locus of the intersection of the ordinate of the ellipse produced with the perpendicular let fall from the centre on the tangent drawn at the point in which the ordinate cuts the ellipse.

20. Find the locus generated by the intersection of two tangents drawn at the extremities of two radii vectores (drawn from centre) which are perpendicular to each other.

$$\text{Ans.} \quad a^4 y^2 + b^4 x^2 = a^2 b^4 + b^2 a^4.$$

21. A line of fixed length so moves that its extremities remain in the co-ordinate axes; required the locus generated by any point of the line.

22. The angle $AOP'' = \varphi$ (Fig. 45) is called the eccentric angle of the point P' (x', y') on the ellipse. Show that $(x', y') = (a \cos \varphi, b \sin \varphi)$ and from these values of the co-ordinates deduce the equation of the ellipse.

23. Express the equation of the tangent at (x'', y'') in terms of the eccentric angle of the point.

$$\text{Ans.} \quad \frac{x}{a} \cos \varphi + \frac{y}{b} \sin \varphi = 1.$$

24. If (x', y'), (x'', y'') are the ends of a pair of conjugate diameters whose eccentric angles are φ and φ', show that $\varphi' - \varphi = 90°$.

CHAPTER VIII.

THE HYPERBOLA.

102. THE hyperbola is the locus of a point so moving in a plane that the *difference* of its distances from two fixed points is always constant and equal to a given line. The fixed points are called the FOCI of the hyperbola. If the points are on the given line produced and equidistant from its extremities, then the given line is called the TRANSVERSE AXIS of the hyperbola.

103. *To deduce the equation of the hyperbola, given the foci and the transverse axis.*

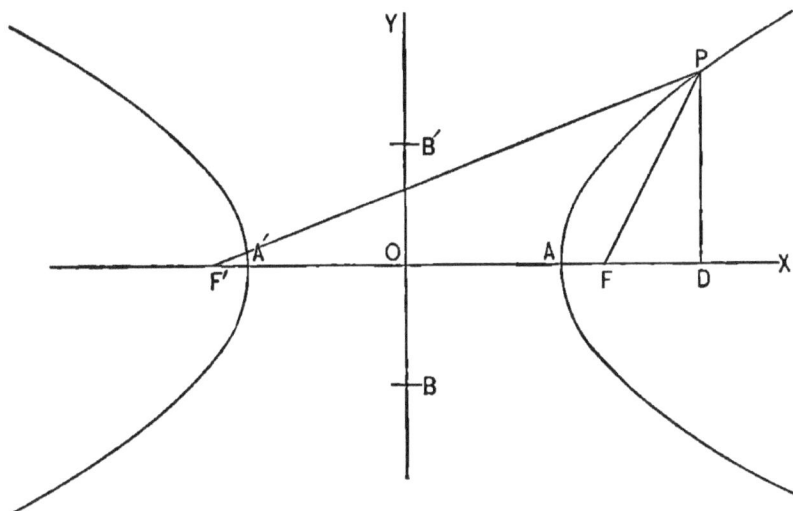

FIG. 47.

Let F, F' be the foci, and AA' the transverse axis. Draw OY ⊥ to AA' at its middle point, and take OY, OX as the co-ordinate axes. Let P be any point of the curve. Draw PF, PF'; draw also PD ∥ to OY.

Then (OD, DP) = (x, y) are the co-ordinates of P.

Let AA' = $2a$, FF' = 2 OF = 2 OF' = $2c$, FP = r and F'P = r'.

From the right angled triangles FPD and F'PD, we have

$$r = \sqrt{y^2 + (x - c)^2} \text{ and } r' = \sqrt{y^2 + (x + c)^2} \cdot \cdot \cdot (a)$$

From the mode of generating the curve, we have

$$r' - r = 2a.$$

Hence, substituting,

$$\sqrt{y^2 + (x + c)^2} - \sqrt{y^2 + (x - c)^2} = 2a; \cdot \cdot \cdot (1)$$

or, clearing of radicals and reducing, we have

$$(c^2 - a^2) x^2 - a^2 y^2 = a^2 (c^2 - a^2) \cdot \cdot \cdot (2)$$

for the required equation. This equation, like that of the ellipse (see Art. 75), may be put in a simpler form.

Let $\qquad c^2 - a^2 = b^2 \cdot \cdot \cdot (3)$

This value in (2) gives, after changing signs,

$$a^2 y^2 - b^2 x^2 = - a^2 b^2, \cdot \cdot \cdot (4)$$

or, symmetrically,

$$\frac{x^2}{a^2} - \frac{y^2}{b^2} = 1 \cdot \cdot \cdot (5)$$

for the equation of the hyperbola when referred to its centre and axes.

Let the student discuss this equation. (See Art. 14) Cor. 1. If $b = a$ in (5), we have

$$x^2 - y^2 = a^2 \cdot \cdot \cdot (6)$$

The curve represented by this equation is called the *Equilateral Hyperbola.* Comparing equation (6) with the equation of the circle

$$x^2 + y^2 = a^2,$$

we see that the equilateral hyperbola bears the same relation to the common hyperbola that the circle bears to the ellipse.

Cor. 2. If (x', y') and (x'', y'') are the co-ordinates of two points on the curve, we have from (4)

$$y'^2 = \frac{b^2}{a^2}(x'^2 - a^2) \text{ and } y''^2 = \frac{b^2}{a^2}(x''^2 - a^2) ;$$

hence $\quad y'^2 : y''^2 :: (x' - a)(x + a) : (x'' - a)(x'' + a) ;$

i.e., *the squares of the ordinates of any two points on the hyperbola are to each other as the rectangles of the segments in which they divide the transverse axis.*

Cor. 3. By making $x = x' - a$ and $y = y'$ in (4) we have after reducing and dropping accents,

$$a^2y^2 - b^2x^2 + 2ab^2x = 0 \ldots (7)$$

for the equation of the hyperbola, A' being taken as origin.

104. From equation (3) Art. 103, we have

$$b = \pm \sqrt{c^2 - a^2}.$$

Laying this distance off above and below the origin on the Y-axis, we have the points B, B', Fig. 47, Art. 103. The line BB' is called the Conjugate Axis of the hyperbola. The points A and A' are called the Vertices of the curve. The point O bisects all lines drawn through it and terminating in the curve; for this reason it is called the Centre of the hyperbola.

The ratio $\quad \dfrac{\sqrt{a^2 + b^2}}{a} = \dfrac{c}{a} = e.\quad$ See (3) Art. 103 $\ldots$ (1)

is called the Eccentricity of the hyperbola. This ratio is evidently > 1. The value of $c = \pm \sqrt{a^2 + b^2}$ measures the distance of the foci F, F' from the centre.

If $b = a$ in (1), we have $e = \sqrt{2}$ for the eccentricity of the equilateral hyperbola.

105. *To find the values of the focal radii, r, r' of a point on the hyperbola in terms of the abscissa of the point.*

From equations (a) Art. 103, we have

$$r = \sqrt{y^2 + (x - c)^2}.$$

From the equation of the hyperbola, (4) Art. 103, we have

$$y^2 = \frac{b^2}{a^2}(x^2 - a^2) = \frac{b^2}{a^2}x^2 - b^2.$$

Hence, substituting

$$r = \sqrt{\frac{b^2}{a^2}x^2 - b^2 + x^2 - 2cx + c^2},$$

$$= \sqrt{\frac{a^2 + b^2}{a^2}x^2 - 2cx + c^2 - b^2},$$

$$= \sqrt{\frac{c^2}{a^2}x^2 - 2cx + a^2}, \quad \text{Art. 104 (1)},$$

$$= \frac{c}{a}x - a ;$$

hence $\qquad r = ex - a \dots (1)$

Similarly, we find

$$r' = ex + a \dots (2)$$

106. *To construct the hyperbola having given the transverse axis and the foci of the curve.*

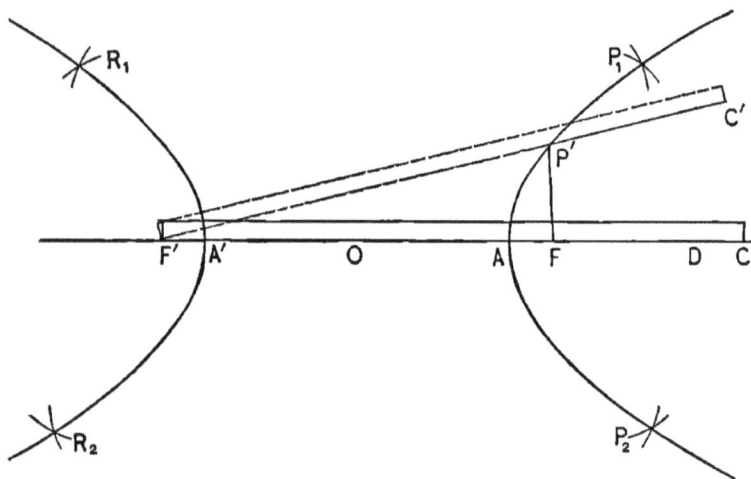

FIG. 48.

First Method. — Let AA′ be the transverse axis and F, F′, the foci. Take a straight-edge ruler whose length is L and attach

one of its ends at F' so that the ruler can freely revolve about
that point. Cut a piece of cord so that its length shall be
= L — 2 a, and attach one end to the free end of the ruler,
and the other end to the focus F. Place the ruler in the
position indicated by the full lines, Fig. 48, and place the
point of a pencil in the loop formed by the cord. Stretch
the cord, keeping the point of the pencil against the edge of
the ruler. If we now revolve the ruler upward about F', the
point of the pencil, kept firmly pressed against the ruler,
will describe the arc AP' of the hyperbola. By fixing the
end of the ruler at F, we may describe an arc of the other
branch. It is evident in this process that the difference of
the distances of the point of the pencil from the foci F',F,
is always equal to 2 a.

Second Method. — Take any point D on the transverse axis.
Measure the distances A'D, AD. With F' as a centre and A'D
as a radius describe the arc of a circle; with F as a centre and
AD as a radius describe another arc. The intersection of
these arcs will determine two points, P_1, P_2, of the curve. By
interchanging centres and radii we may locate the points R_1,
R_2, on the other branch. In this manner we may determine as
many points as the accuracy of the construction may require.

107. *To find the latus-rectum or parameter of the hyperbola.*
The LATUS-RECTUM, *or* PARAMETER *of the hyperbola, is the
double ordinate passing through either focus.*
Making $x = \pm \sqrt{a^2 + b^2}$ in the equation of the hyperbola

$$y^2 = \frac{b^2}{a^2} (x^2 - a^2),$$

we have $\qquad y = \frac{b^2}{a} \therefore 2\, y = \frac{2\, b^2}{a}.$

Forming a proportion from this equation, we have
$$2\, y : 2\, b :: b : a;$$
$$\therefore 2\, y : 2\, b :: 2\, b : 2\, a;$$

i.e, *the latus-rectum of the hyperbola is a third proportional to
the axes.*

108. The equation of the ellipse when referred to its centre and axes is

$$a^2y^2 + b^2x^2 = a^2b^2.$$

The equation of the hyperbola when referred to its centre and axes is

$$a^2y^2 - b^2x^2 = - a^2b^2.$$

Comparing these equations, we see that the only difference is in the sign of b^2. If, therefore, in the various analytical expressions we have deduced for the ellipse, we substitute $- b^2$ for b^2, or, what is the same thing, $+ b \sqrt{- 1}$ for b, we will obtain the corresponding analytical expressions for the hyperbola.

109. *To deduce the equation of the conjugate hyperbola. Two hyperbolas are* Conjugate *when the transverse and conjugate axes of one are respectively the conjugate and transverse axes of the other.*

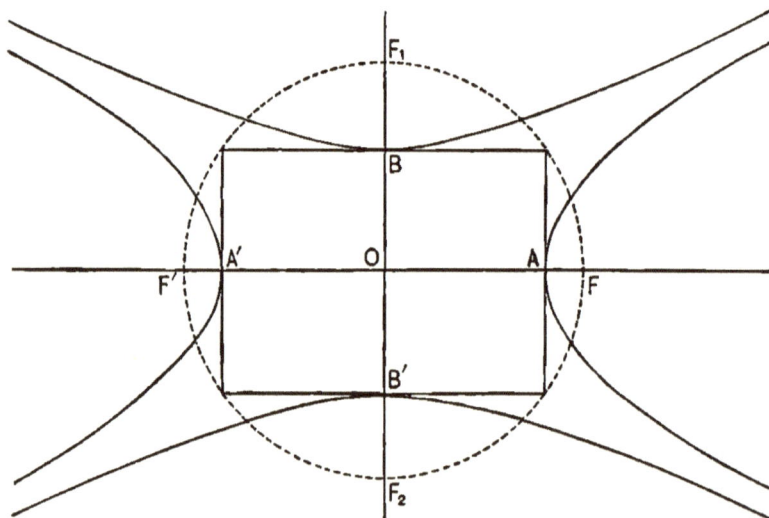

Fig. 49.

Thus in Fig. 49, if AA' be the transverse axis of the hyperbola which has BB' for its conjugate axis, then the hyperbola which has BB' for its transverse axis and AA' for its conjugate

axis is its conjugate; and, conversely, the hyperbola whose
transverse axis is BB′ and conjugate axis is AA′ has for its
conjugate the hyperbola whose transverse axis is AA′ and
whose conjugate axis is BB′.

We have deduced, Art. 103, (5),

$$\frac{x^2}{a^2} - \frac{y^2}{b^2} = 1 \ldots (1)$$

for the equation of the hyperbola whose transverse axis lies
along the X-axis. We wish to find the equation of its conjugate. It is obvious from the figure that the hyperbola which
has BB′ for its transverse axis and AA′ for its conjugate axis
bears the same relation to the Y-axis as the hyperbola whose
transverse axis is AA′ and conjugate axis is BB′ bears to the
X-axis; hence, changing a to b and b to a, x to y and y to x
in (1), we have

$$\frac{y^2}{b^2} - \frac{x^2}{a^2} = 1,$$

or
$$\frac{x^2}{a^2} - \frac{y^2}{b^2} = -1 \ldots (2)$$

for the equation of the conjugate hyperbola to the hyperbola
whose equation is (1).

Comparing (1) and (2) we see that *the equation of any
hyperbola and that of its conjugate differ only in the sign of
the constant term.*

Cor. Since $\sqrt{b^2 + a^2} = \sqrt{a^2 + b^2}$, the focal distances of
any hyperbola and those of its conjugate are equal.

The eccentricities of conjugate hyperbolas, however, are
not equal. For the hyperbola whose semi-transverse axis is
a and semi-conjugate axis is b, we have

Art. 104, (1) $e = \dfrac{\sqrt{a^2 + b^2}}{a}$.

For its conjugate hyperbola, we have

$$e' = \frac{\sqrt{a^2 + b^2}}{b}.$$

EXAMPLES.

Find the semi-axes, the eccentricity and the latus-rectum of each of the following hyperbolas :

1. $9\,y^2 - 4\,x^2 = -36.$ **5.** $3\,y^2 - 2\,x^2 = 12.$

2. $\dfrac{x^2}{4} - \dfrac{y^2}{9} = 1.$ **6.** $ay^2 - bx^2 = -ab.$

3. $y^2 - 16\,x^2 = -16.$ **7.** $\dfrac{x^2}{4} - y^2 = m.$

4. $4\,x^2 - 16\,y^2 = -64.$ **8.** $y^2 - mx^2 = n.$

Write the equation of the hyperbola having given:

9. The transverse axis $= 12$; the distance between the foci $= 16.$

$$\textit{Ans.} \quad \frac{x^2}{36} - \frac{y^2}{28} = 1.$$

10. The transverse axis $= 10$; parameter $= 8.$

$$\textit{Ans.} \quad \frac{x^2}{25} - \frac{y^2}{20} = 1.$$

11. Semi-conjugate axis $= 6$; the focal distance $= 10.$

$$\textit{Ans.} \quad \frac{x^2}{64} - \frac{y^2}{36} = 1.$$

12. The equation of the conjugate hyperbola to $x^2 - 3\,y^2 = 6.$

$$\textit{Ans.} \quad x^2 - 3\,y^2 + 6 = 0.$$

13. The conjugate axis is 10, and the transverse axis is double the conjugate.

$$\textit{Ans.} \quad \frac{x^2}{100} - \frac{y^2}{25} = 1.$$

14. The transverse axis is 8, and the conjugate axis $= \tfrac{1}{2}$ distance between foci.

$$\textit{Ans.} \quad \frac{x^2}{16} - \frac{3y^2}{16} = 1.$$

15. Given the hyperbola

$$\frac{x^2}{10} - \frac{y^2}{4} = 1 ;$$

required the co-ordinates of the point whose abscissa is double its ordinate.

$$Ans. \quad \left(2\sqrt{\frac{20}{3}}, \ \sqrt{\frac{20}{3}} \right).$$

16. Write the equation of the conjugate hyperbola to each of the hyperbolas given in the first eight examples above.

17. Given the hyperbola $9\,y^2 - 4\,x^2 = -36$; required the focal radii of the point whose ordinate is $= 1$ and abscissa positive.

18. Determine the points of intersection of

$$\frac{x^2}{4} - \frac{y^2}{9} = 1, \text{ and } \frac{x^2}{16} + \frac{y^2}{16} = 1.$$

110. *To deduce the polar equation of the hyperbola, either focus being taken as the pole.*

Let us take F as the pole, Fig. 47.

Let $(FP, PFD) = (r, \theta)$ be the co-ordinates of any point P on the curve. From Art. 105, (1), we have

$$FP = r = ex - a \ . \ . \ . \ (1)$$

From Fig. 47, $OD = OF + FD$;

i.e., $x = ae + r \cos \theta.$

Substituting this value in (1) and reducing, we have

$$r = -\frac{a\,(1 - e^2)}{1 - e \cos \theta} \ . \ . \ . \ (2)$$

for the polar equation of the hyperbola, the right hand focus being taken as the pole.

Similarly from Art. 105, (2), we have

$$r' = \frac{a\,(1 - e^2)}{1 - e \cos \theta} \ . \ . \ . \ (3)$$

for the polar equation, the left hand focus being the pole.

Cor.   If $\theta = 0$, $r = -a - ae = -$ FA′,
$$r' = a + ae = \text{F}'\text{A}.$$

If $\theta = 90°$,
$$r = -a + ae^2 = \frac{a^2e^2 - a^2}{a} = \frac{b^2}{a} = \textit{semi-latus rectum}.$$
$$r' = a - ae^2 = \frac{a^2 - a^2e^2}{a} = -\frac{b^2}{a} = \textit{semi-latus rectum}.$$

If $\theta = 180°$, $r = -a + ae = $ FA,
$$r' = a - ae = -\text{F}'\text{A}'.$$

If $\theta = 270°$, $r = -a + ae^2 = \frac{b^2}{a} = \textit{semi-latus rectum}.$
$$r' = a - ae^2 = -\frac{b^2}{a} = \textit{semi-latus rectum}.$$

111. *To deduce the equation of condition for the supplementary chords of the hyperbola.*

By a method similar to that of **Art. 81**, or by placing $-b^2$ for b^2 in (3) of that article, we have
$$ss' = \frac{b^2}{a^2}; \; \ldots \; (1)$$

hence, *the product of the slopes of any pair of supplementary chords of an hyperbola is the same for every pair.*

Cor.   If $a = b$, we have
$$ss' = 1, \text{ or, } s = \frac{1}{s'},$$
$$\therefore \tan \alpha = \cot \alpha';$$

hence, *the sum of the two acute angles which any pair of supplementary chords of an equilateral hyperbola make with the X-axis is equal to* 90°.

112. *To deduce the equation of the tangent to the hyperbola.*

By a method entirely analogous to that adopted in the circle, or ellipse, or parabola, **Arts. 41, 82, 57**; or substituting $-b^2$ for b^2 in (5) of **Art. 82**, we find
$$\frac{xx''}{a^2} - \frac{yy''}{b^2} = 1 \; \ldots \; (1)$$

to be *the equation of the tangent to the hyperbola.*

113. *To deduce the value of the sub-tangent.*

By operating on (1) of the preceding article (see Art. 83), we find

$$\text{Sub-tangent} = x'' - \frac{a^2}{x''} = \frac{x''^2 - a^2}{x''}.$$

114. The slope of a line passing through the centre of an hyperbola (0, 0) and the point of tangency (x'', y'') is

$$t = \frac{y''}{x''}.$$

The slope of the tangent is, Art. 112, (1)

$$t' = \frac{b^2}{a^2} \cdot \frac{x''}{y''}.$$

Multiplying these equations, member by member, we have

$$tt' = \frac{b^2}{a^2} \ldots (1)$$

Comparing (1) of this article with (1) of Art. 111, we find

$$ss' = tt' \ldots (2)$$

Hence, the line from the centre of the hyperbola to the point of tangency and the tangent enjoy the property of being the supplemental chords of an hyperbola whose semi-axes bear to each other the ratio $\frac{b}{a}$.

COR. If $s = t$, then $s' = t'$; i.e., if one supplementary chord of an hyperbola is parallel to a line drawn through the centre, then the other supplementary chord is parallel to the tangent drawn to the curve at the point in which the line through the centre cuts the curve.

115. The preceding principle affords us a simple method of drawing a tangent to the hyperbola at any given point of the curve.

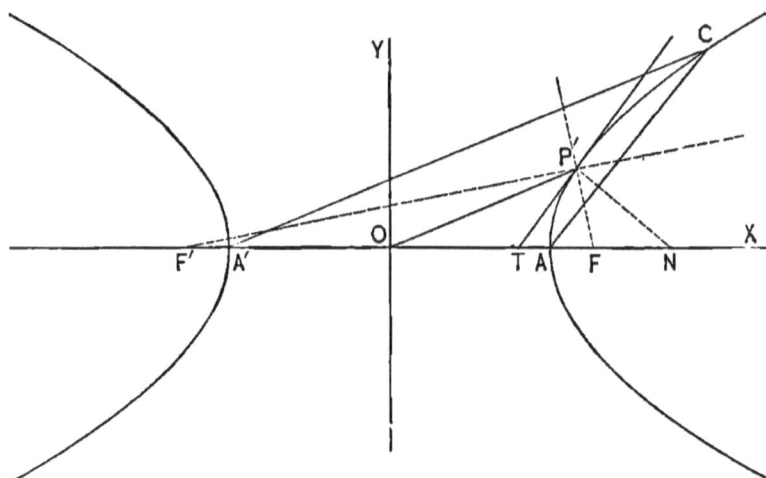

FIG. 50.

Let P′ be any point at which we wish to draw a tangent. Join P′ and O, and from A′ draw A′C ∥ to P′O; join C and A. The line P′T, drawn from P″ ∥ to CA will be the required tangent.

116. *To deduce the equation of the normal to the hyperbola.*

We can do this by operating on the equation of the tangent, as in previous cases, or by changing b^2 into $-b^2$ in the equation of the normal to the ellipse, Art. 86, (3). By either method, we obtain

$$y - y'' = -\frac{a^2 y''}{b^2 x''}(x - x'') \ \ldots \ (1)$$

for the required equation.

117. *To deduce the value of the sub-normal.*

By a course of reasoning similar to that of Art. 87, we have

$$sub\text{-}normal = \frac{b^2}{a^2}x''.$$

Cor. If $b = a$,

$$sub\text{-}normal = x'';$$

i.e., in the equilateral hyperbola the sub-normal is equal to the abscissa of the point of tangency.

EXAMPLES.

1. Deduce the polar equation of the hyperbola, the pole being at the centre.

$$r^2 = -\frac{a^2 b^2}{a^2 \sin^2 \theta - b^2 \cos^2 \theta}.$$

Write the equation of the tangents to each of the following hyperbolas, and give the value of the sub-tangent in each case.

2. $9 y^2 - 4 x^2 = -36$, at (4, ord. +).

3. $\dfrac{y^2}{9} - \dfrac{x^2}{16} = -1$, at (5, ord. +).

4. $\dfrac{x^2}{9} - \dfrac{y^2}{16} = 1$, at (4, ord. +).

5. $y^2 - 4 x^2 = -36$, at (abs. +, 6).

6. $a y^2 - b x^2 = -ab$, at ($\sqrt{ab}$, ord. +).

7. $\dfrac{x^2}{m} - \dfrac{y^2}{n} = 1$, at ($\sqrt{m}$, 0).

8. Write the equation of the normal to each of the above hyperbolas, and give the value of the sub-normal in each case.

9. The equation of a chord of an hyperbola is $y - x - 6 = 0$; what is the equation of the supplemental chord, the axes of the hyperbola being 12 and 8 ?

$$Ans. \quad y = \frac{4}{9} x - \frac{8}{3}.$$

10. Given the equations

$$\frac{x^2}{9} - \frac{y^2}{4} = -1, \text{ and } y - x = 0 ;$$

required the equations of the tangents to the hyperbola at the points in which the line pierces the curve.

11. One of the supplementary chords of the hyperbola $9\,y^2 - 16\,x^2 = -\,144$ is parallel to the line $y = x$; what are the equations of the chords ?

$$Ans. \quad \begin{cases} y = x + 3 \\ y = \dfrac{16}{9}x - \dfrac{16}{3} \end{cases}$$

12. Given the hyperbola $2\,x^2 - 3\,y^2 = 6$; required the equations of the tangent and normal at the positive end of the right hand focal ordinate.

13. What is the equation of a tangent to

$$\frac{x^2}{4} - \frac{y^2}{6} = 1,$$

which is parallel to the line $2\,y - x + 1 = 0$?

118. *The angle formed by the focal lines drawn to any point of the hyperbola is bisected by the tangent at that point.*

Making $y = o$ in the equation of the tangent line, Art. 112, (1), we have

$$x = \frac{a^2}{x''} = \mathrm{OT}. \quad \text{Fig. 50.}$$

From Art. 104, (1) $\mathrm{OF} = \mathrm{OF'} = ae$;

hence $\quad \mathrm{OF} - \mathrm{OT} = \mathrm{FT} = ae - \dfrac{a^2}{x''} = \dfrac{a}{x''}\,(ex'' - a).$

$$\mathrm{OF'} + \mathrm{OT} = \mathrm{F'T} = ae + \frac{a^2}{x''} = \frac{a}{x''}\,(ex'' + a);$$

$$\therefore \mathrm{FT} : \mathrm{F'T} :: ex'' - a : ex'' + a.$$

But from Art. 105 we have

$$\mathrm{FP'} = ex'' - a$$
$$\mathrm{F'P'} = ex'' + a;$$
$$\therefore \mathrm{FP'} : \mathrm{F'P'} :: ex'' - a : ex'' + a.$$

Hence $\quad \mathrm{FT} : \mathrm{F'T} :: \mathrm{FP'} : \mathrm{F'P'};$

i.e., the tangent P'T divides the base of the triangle FP'F' into two segments, which are proportional to the adjacent sides; it must therefore bisect the angle at the vertex.

Cor. Since the normal P'N, Fig. 50, is perpendicular to the tangent, it bisects the external angle formed by the focal radii.

Schol. The principle of this article gives us another method of drawing a tangent to the hyperbola at a given point. Let P' be the point, Fig. 50. Draw the focal radii FP', F'P'. The line P'T drawn so as to bisect the angle between the focal radii will be tangent to the curve at P'.

119. *To find the condition that the line* $y = sx + c$ *must fulfil in order that it may touch the hyperbola*

$$\frac{x^2}{a^2} - \frac{y^2}{b^2} = 1.$$

By a method similar to that employed in Art. 89, we find

$$s^2 a^2 - b^2 = c^2 \ . \ . \ . \ (1)$$

for the required condition.

Cor. 1. Substituting the value of c drawn from (1) in the equation of the line, we have

$$y = sx \pm \sqrt{s^2 a^2 - b^2} \ . \ . \ . \ (2)$$

for *the equation of the tangent to the hyperbola in terms of its slope.*

120. *To find the locus generated by the intersection of a tangent to the hyperbola and a perpendicular to it from a focus as the point of tangency moves around the curve.*

$$x^2 + y^2 = a^2 \ . \ . \ . \ (1)$$

is the equation of the required locus. (See Art. 90.)

121. *To find the locus generated by the intersection of two tangents which are perpendicular to each other as the points of tangency move around the curve.*

$$x^2 + y^2 = a^2 - b^2 \ . \ . \ . \ (1)$$

is the equation of the required locus. (See Art. 91.)

122. *Two tangents are drawn to the hyperbola from a point without ; required the equation of the line joining the points of tangency.*

$$\frac{x'x}{a^2} - \frac{y'y}{b^2} = 1 \ldots (1)$$

is the required equation.   (See Art. 92.)

123. *To find the equation of the polar of the pole (x', y'), with regard to the hyperbola*

$$\frac{x^2}{a^2} - \frac{y^2}{b^2} = 1.$$

$$\frac{x'x}{a^2} - \frac{y'y}{b^2} = 1 \ldots (1)$$

is the required equation.   (See Arts. 49 and 93.)

124. *To deduce the equation of the hyperbola when referred to a pair of conjugate diameters.*

A pair of diameters are said to be conjugate when they are so related that the equation of the hyperbola, when the curve is referred to them as axes, contains only the second powers of the variables.

$$\frac{x^2}{a'^2} - \frac{y^2}{b'^2} = 1 \ldots (1)$$

is the required equation, and

$$a^2 \sin \theta \sin \varphi - b^2 \cos \theta \cos \varphi = 0,$$

or $$\tan \theta \tan \varphi = \frac{b^2}{a^2} \ldots (2)$$

is *the condition for conjugate diameters.*   (See Art. 94.)

Cor. From the form of (1) we see that *all chords drawn parallel to one of two conjugate diameters are bisected by the other.*

Schol. From Art. 111, (1) we have

$$ss' = \frac{b^2}{a^2};$$

hence $ss' = \tan \theta \tan \varphi.$

If, therefore, $s = \tan \theta$, we have $s' = \tan \varphi$; i.e., *if one of two conjugate diameters is parallel to a chord, the other conjugate diameter is parallel to the supplement of that chord.*

From Art. 114 we have

$$tt' = \frac{b^2}{a^2};$$

hence $\qquad tt' = \tan \theta \tan \varphi.$

If, therefore, $t = \tan \theta$, we have $t' = \tan \varphi$; i.e., *if one of two conjugate diameters is parallel to a tangent of the hyperbola, the other conjugate diameter coincides with the line joining the point of tangency and the centre.*

125. From the condition for conjugate diameters,

$$\tan \theta \tan \varphi = \frac{b^2}{a^2},$$

we see that the products of the slopes of any pair of conjugate diameters is positive; hence, the slopes are both *positive* or both *negative.* It appears, therefore, *that any two conjugate diameters must lie in the same quadrant.*

126. *To find the equation of a conjugate diameter.*

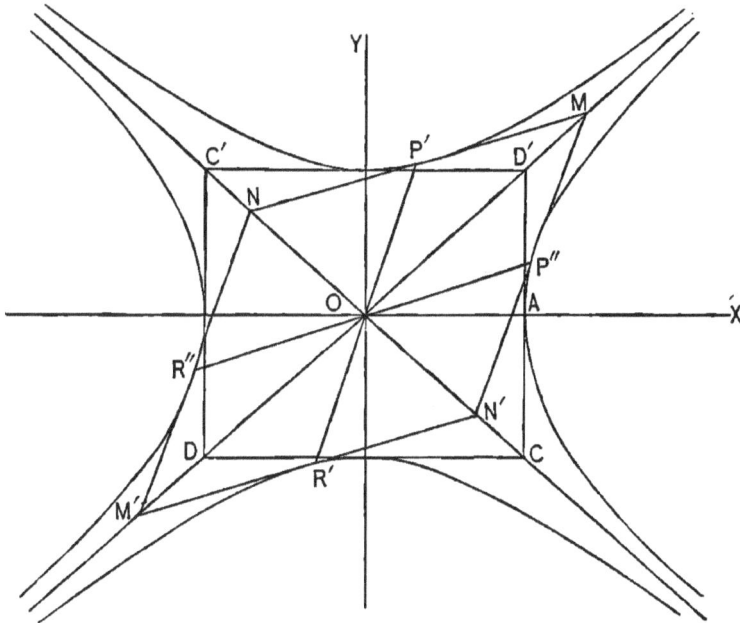

FIG. 51.

Let $P''R''$ be *any* diameter; then $P'R'$, drawn through the centre O parallel to the tangent at P'' ($P''N'$) will be its conjugate diameter. Art. 124, Schol.

The equation of the tangent at P'' (x'', y'') is

$$\frac{xx''}{a^2} - \frac{yy''}{b^2} = 1 \ldots (1)$$

hence, the equation of $P'R'$ is

$$\frac{xx''}{a^2} - \frac{yy''}{b^2} = 0,$$

or

$$y = \frac{b^2}{a^2} \cdot \frac{x''}{y''} x \ldots (2)$$

But

$$\frac{x''}{y''} = \cot P''OX = \frac{1}{s};$$

hence

$$y = \frac{b^2}{a^2 s} x \ldots (3)$$

is the equation of a diameter in terms of the slope of its conjugate diameter.

127. *To find the co-ordinates of either extremity of a diameter, the co-ordinates of one extremity of its conjugate diameter being given.*

Let the co-ordinates of P'' (x'', y''), Fig. 51, be given. By a course of reasoning similar to that of Art. 96, we find

$$x' = \pm \frac{a}{b} y'', y' = \pm \frac{b}{a} x''.$$

The upper signs correspond to the point P' (x', y'); the lower signs to the point R' $(-x', -y')$.

128. *To show that the difference of the squares of any pair of semi-conjugate diameters is equal to the difference of the squares of the semi-axes.*

By a course of reasoning similar to that of Art. 97, or, by substituting $-b^2$ for b^2, $-b'^2$ for b'^2 in (4) of that article, we find

$$a'^2 - b'^2 = a^2 - b^2 \ldots (1)$$

Cor. If $a = b$, then $a' = b'$; i.e., *the equilateral hyperbola has equal conjugate diameters.*

129. *To show that the parallelogram constructed on any two conjugate diameters is equivalent to the rectangle constructed on the axes.*

By a method similar to that of Art. 98, we can show that

$$4\, a'b' \sin{(\varphi - \theta)} = 4\, ab;$$

i.e., Area $\text{MNM'N'} = $ Area CDC'D'. Fig. 51.

EXAMPLES.

1. The line $y = 2x + c$ touches the hyperbola

$$\frac{x^2}{9} - \frac{y^2}{4} = 1;$$

what is the value of c?

Ans. $c = \pm \sqrt{32}$.

2. A tangent to the hyperbola

$$\frac{x^2}{10} - \frac{y^2}{12} = 1$$

has its Y-intercept $= 2$; required its slope and equation.

Ans. $\sqrt{1.6}$; $y = \sqrt{1.6}\, x + 2$.

3. A tangent to the hyperbola $4y^2 - 2x^2 = 6$ makes an angle of $45°$ with the X-axis; required its equation.

4. Two tangents are drawn to the hyperbola $4y^2 - 9x^2 = -36$ from the point $(1, 2)$; required the equation of the chord of contact.

Ans. $9x - 8y = 36$.

5. What is the equation of the polar of the right-hand focus? Of the left-hand focus?

6. What is the polar of $(1, \frac{1}{2})$ with regard to the hyperbola $4y^2 - x^2 = -4$? *Ans.* $x - 2y = 4$.

7. Find the diameter conjugate to $y = x$ in the hyperbola

$$\frac{x^2}{9} - \frac{y^2}{16} = 1.$$

Ans. $y = \tfrac{16}{9} x$.

8. Given the chord $y = 2x + 6$ of the hyperbola

$$\frac{x^2}{9} - \frac{y^2}{4} = 1;$$

required the equations of the supplementary chord.

Ans. $y = \frac{2}{3}x - \frac{2}{3}$.

9. In the last example find the equation of the pair of conjugate diameters which are parallel to the chords.

Ans. $y = 2x, 9y = 2x$.

10. The point $(5, \frac{16}{3})$ lies on the hyperbola $9y^2 - 16x^2 = -144$; required the equation of the diameter passing through it; also the co-ordinates of the extremities of its conjugate diameter.

130. *To deduce the equations of the rectilinear asymptotes of the hyperbola.*

An ASYMPTOTE *of a curve is a line passing within a finite distance of the origin which the curve continually approaches, and to which it becomes tangent at an infinite distance.*

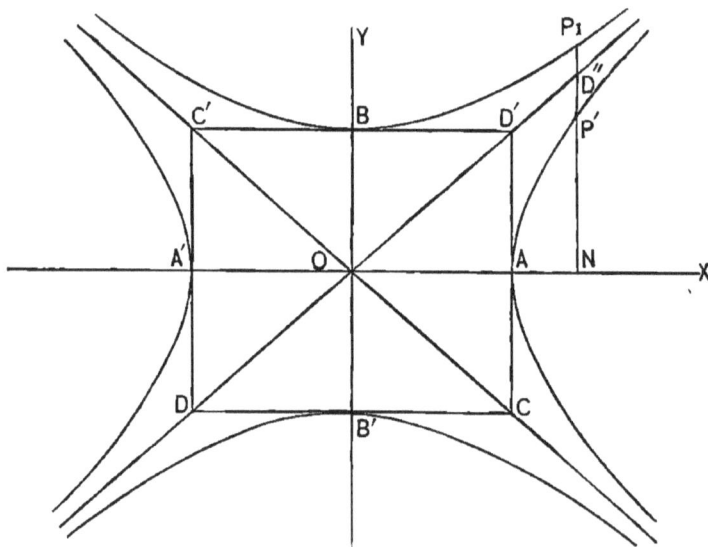

FIG. 52.

The equation of the hyperbola whose transverse axis lies along the X-axis may be put under the form

$$y^2 = \frac{b^2}{a^2}(x^2 - a^2) \ \ldots \ (1)$$

The equations of the diagonals, DD', CC', of the rectangle constructed on the axes AA', BB' are

$$y' = \pm \frac{b}{a}x,$$

or, squaring, $\quad y'^2 = \dfrac{b^2}{a^2}x^2 \ \ldots \ (2)$

where y' represents the ordinates of points on the diagonals.

Let P' (x, y) be any point on the X-hyperbola; and let D'' (x, y') be the corresponding point on the diagonal DD'. Subtracting (1) from (2) and factoring, we have

$$(y' - y)(y' + y) = b^2;$$

hence $\quad y' - y = \text{D''P'} = \dfrac{b^2}{y' + y} \ \ldots \ (3)$

As the points D'', P' recede from the centre, O, their ordinates D''N, P'N increase and become infinite in value when D'' and P' are at an infinite distance. But as the ordinates increase the value of the fraction (3), which represents their difference, decreases and becomes zero when y' and y are infinite; hence, the points D'' and P' are continually approaching each other as they recede from the centre until at infinity they coincide. But the locus of D'' during this motion is the infinite diagonal DD'; hence, *the diagonals of the rectangle constructed on the axes of the hyperbola are the asymptotes of the curve.*

Therefore $\quad y = + \dfrac{b}{a}x$ and $y = -\dfrac{b}{a}x$

are the required equations.

Cor. 1. If $a = b$, then

$$y = + x \text{ and } y = - x;$$

i.e., *the asymptotes of the equilateral hyperbola make angles of 45° with the X-axis.*

COR. 2. The equation of the hyperbola conjugate to (1) may be put under the form

$$y''^2 = \frac{b^2}{a^2}(x^2 + a^2) \ldots (4)$$

Subtracting (1) from (4), we have

$$y'' - y = P_1 P' = \frac{2b^2}{y'' + y};$$

hence, *an hyperbola and its conjugate are curvilinear asymptotes of each other.*

COR. 3. Subtracting (2) from (4), we have

$$y'' - y' = P_1 D'' = \frac{b^2}{y'' + y'};$$

hence, *the rectilinear asymptotes of an hyperbola and of its conjugate are the same.*

131. *To deduce the equation of the hyperbola when referred to its rectilinear asymptotes as axes.*

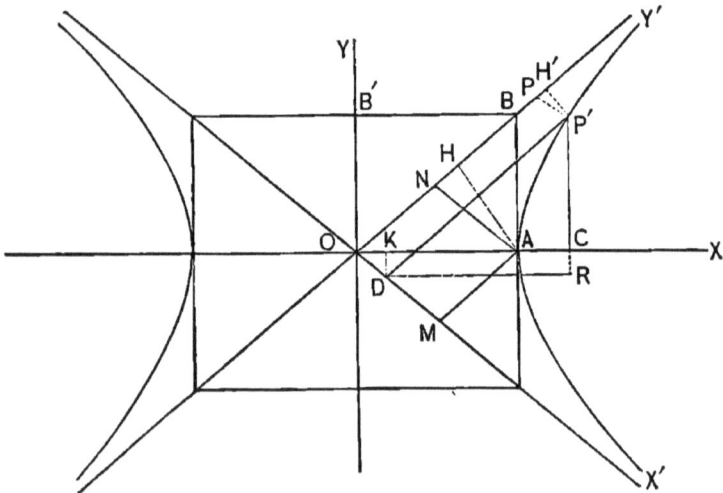

FIG. 53.

The equation of the hyperbola when referred to OY, OX, is

$$\frac{x^2}{a^2} - \frac{y^2}{b^2} = 1 \ldots (1)$$

We wish to ascertain what this equation becomes when OY', OX' the rectilinear asymptotes are taken as axes.

Let P' be any point of the curve; let $Y'OX = XOX' = \theta$. Then $(OC, CP') = (x, y)$; $(OD, DP') = (x', y')$.

From the figure, $OC = OK + DR$; $CP' = RP' - DK$;

i.e., $\qquad x = (x' + y') \cos \theta$; $y = (y' - x') \sin \theta$.

But from the triangle OAB, we have

$$\sin \theta = \frac{AB}{OB} = \frac{b}{\sqrt{a^2 + b^2}},$$

$$\cos \theta = \frac{OA}{OB} = \frac{a}{\sqrt{a^2 + b^2}};$$

hence, $x = (x' + y') \dfrac{a}{\sqrt{a^2 + b^2}}$; $y = (y' - x') \dfrac{b}{\sqrt{a^2 + b^2}}$.

Substituting these values in (1), we have

$$(x' + y')^2 - (y' - x')^2 = a^2 + b^2,$$

or, reducing and dropping accents,

$$xy = \frac{a^2 + b^2}{4} \quad \ldots \quad (2)$$

for the equation of the hyperbola referred to its asymptotes.

In a similar manner we may show that

$$xy = - \frac{a^2 + b^2}{4} \quad \ldots \quad (3)$$

is the equation of the hyperbola conjugate to (1), when referred to its asymptotes as axes.

COR. Multiplying (2) by $\sin 2\theta$ we may place the result in the form

$$yx \sin 2\theta = \frac{\sqrt{a^2 + b^2}}{2} \cdot \frac{\sqrt{a^2 + b^2}}{2} \sin 2\theta;$$

that is $\qquad$ DP'. P'H' = ON. AH;

therefore $\qquad$ *area* ODP'P = *area* OMAN;

hence, *the area of the parallelogram constructed upon the co-ordinates of any point of the hyperbola, the asymptotes being axes, is constant and equal to the area of the rhombus constructed upon the co-ordinates of the vertex.*

132. *To deduce the equation of the tangent to the hyperbola when the curve is referred to its rectilinear asymptotes as axes.*

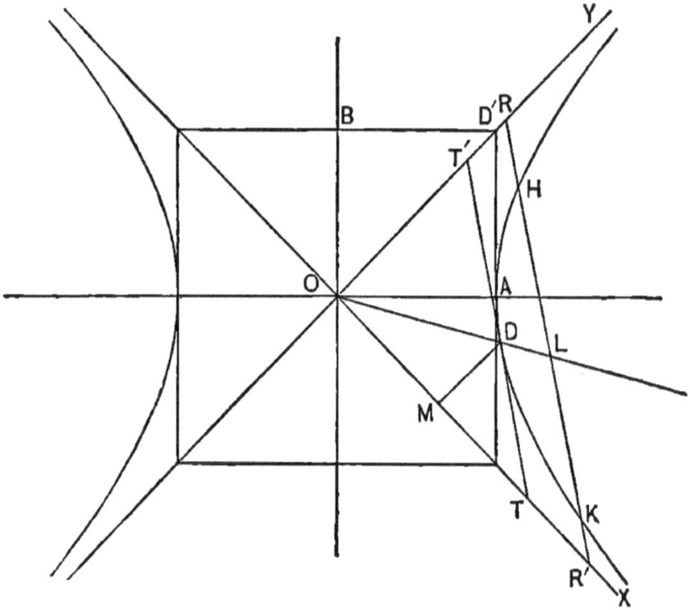

FIG. 54.

By a course of reasoning similar to that employed in **Arts.** 41, 57, 82, we find the required equation to be

$$y - y'' = -\frac{y''}{x''}(x - x'') \ \cdots \ (1)$$

or, symmetrically,

$$\frac{x}{x''} + \frac{y}{y''} = 2 \ \cdots \ (2)$$

Cor. If we make $y = 0$ in (2), we have

$$x = 2x'' = OT. \quad \text{Fig. 54.}$$

But $\quad OM = x'', \therefore OM = MT \therefore T'D = TD;$

hence, *the point of tangency in the hyperbola bisects that portion of the tangent included between the asymptotes.*

133. Since D (x'', y'') is a point of the hyperbola, we have (see Fig. 54)

$$4\,x''y'' = a^2 + b^2,$$

or $$2\,x'' \cdot 2\,y'' = a^2 + b^2;$$

i.e., $$OT \cdot OT' = a^2 + b^2 \ldots (1)$$

hence, *the rectangle of the intercepts of a tangent on the asymptotes is constant and equal to the sum of the squares on the semi-axes.*

134. From (1) of the last article we have, after multiplying through by $\dfrac{\sin 2\,\theta}{2}$,

$$\frac{OT \cdot OT'}{2} \sin 2\,\theta = \frac{a^2 + b^2}{2} \sin 2\,\theta = (a^2 + b^2)\sin \theta \cos \theta.$$

But, Art. 131,

$$\sin \theta = \frac{b}{\sqrt{a^2 + b^2}}, \quad \cos \theta = \frac{a}{\sqrt{a^2 + b^2}};$$

hence $$\frac{OT \cdot OT'}{2} \sin 2\,\theta = ab;$$

i.e., $$area\ OTT' = area\ OAD'B.$$

$\therefore$ *the triangle formed by a tangent to the hyperbola and its asymptotes is equivalent to the rectangle on the semi-axes.*

135. Draw the chord RR', Fig. 54, parallel to the tangent T'T. Draw also the diameter OL through D.

Since TD = T'D, we have R'L = RL.

Since OL is a diameter, we have LK = LH; hence

$$R'L - LK = RL - LH;$$

i.e., $$R'K = RH;$$

hence, *the intercepts of a chord between the hyperbola and its asymptotes are equal.*

EXAMPLES.

1. What are the equations of the asymptotes of the hyperbola $\dfrac{x^2}{9} - \dfrac{y^2}{16} = 1$?

Ans. $y = \pm \tfrac{4}{3} x.$

What are the equations of the asymptotes of the following hyperbolas:

2. $\dfrac{x^2}{16} - y^2 = 1.$ **4.** $\dfrac{y^2}{10} - \dfrac{x^2}{9} = 1.$

Ans. $y = \pm \dfrac{x}{4}.$

3. $3 y^2 - 2 x^2 = -6.$ **5.** $m x^2 - n y^2 = c.$

Ans. $y = \pm \sqrt{\tfrac{2}{3}}\, x.$

6. What do the equations given in the four preceding examples become when the hyperbolas which they represent are referred to their asymptotes as axes?

7. The semi-conjugate axis of the hyperbola $xy = 25$ is 6; what is the value of the semi-transverse axis?

Ans. 8.

What are the equations of the tangents to the following hyperbolas:

8. To $xy = 10$, at $(1, 10)$.

Ans. $y + 10 x = 20.$

9. To $xy = +12$, at $(2, 6)$.

Ans. $y = -3 x + 12.$

10. To $xy = m$, at $(-1, -m)$.

11. To $xy = -p$, at $\left(-2, \dfrac{p}{2}\right).$

12. Required the point of the hyperbola $xy = 12$ for which the sub-tangent $= 4$.

Ans. $(4, 3).$

13. The equations of the asymptotes of an hyperbola whose transverse axis $= 16$ are $3 y = 2 x$ and $3 y + 2 x = 0$; required the equation of the hyperbola.

Ans. $\dfrac{x^2}{64} - \dfrac{9 y^2}{256} = 1.$

14. Prove that the product of the perpendiculars let fall from any point of the hyperbola on the asymptotes is constant and

$$= \frac{a^2 b^2}{a^2 + b^2}.$$

GENERAL EXAMPLES.

1. The point (6, 4) is on the hyperbola whose transverse is 10; required the equation of the hyperbola.

Ans. $\dfrac{x^2}{25} - \dfrac{11\,y^2}{400} = 1.$

2. Assume the equation of the hyperbola, and show that the difference of the distances of any point on it from the foci is constant and $= 2\,a$.

3. Required the equation of the hyperbola, transverse axis $= 6$, which has $5\,y = 2\,x$ and $3\,y = 13\,x$ for the equations of a pair of conjugate diameters.

Ans. $\dfrac{x^2}{9} - \dfrac{5\,y^2}{78} = 1.$

4. Show that the ratio of the sum of the focal radii of any point on the hyperbola to the abscissa of the point is constant and $= 2\,e$.

5. What are the conditions that the line $y = sx + c$ must fulfil in order to touch

$$\frac{x^2}{a^2} - \frac{y^2}{b^2} = 1 \text{ at infinity?}$$

Ans. $s = \pm \dfrac{b}{a},\ c = 0.$

6. Show that the conjugate diameters of an hyperbola are also the conjugate diameters of the conjugate hyperbola.

7. Show that the portions of the chord of an hyperbola included between the hyperbola and its conjugate are equal.

8. What is the equation of the line which passes through the focus of an hyperbola and the focus of its conjugate hyperbola?

Ans. $x + y = \sqrt{a^2 + b^2}.$

9. Show that

$$\frac{e}{e'} = \frac{b}{a}$$

when e and e' are the eccentricities of two conjugate hyperbolas.

10. Find the angle between any pair of conjugate diameters of the hyperbola.

11. Show that in the hyperbola the curve can be cut by only one of two conjugate diameters.

12. Find whether the line $y = \frac{4}{3}x$ intersects the hyperbola $16\,y^2 - 9\,x^2 = -144$, or its conjugate.

13. Show that the conjugate diameters of the equilateral hyperbola make equal angles with the asymptotes.

14. Show that lines drawn from any point of the equilateral hyperbola to the extremities of a diameter make equal angles with the asymptotes.

15. In the equilateral hyperbola focal chords drawn parallel to conjugate diameters are equal.

16. A perpendicular is drawn from the focus of an hyperbola to the asymptote : show

(*a*) that the foot of the perpendicular is at the distance a from the centre, and

(*b*) that the foot of the perpendicular is at the distance b from the focus.

17. For what point of an hyperbola is the sub-tangent = the sub-normal ?

18. Show that in the equilateral hyperbola the length of the normal is equal to the distance of the point of contact from the centre.

19. Show that the tangents drawn at the extremities of any chord of the hyperbola intersect on the diameter which bisects the chord.

20. Find the equation of the chord of the hyperbola

$$\frac{x^2}{9} - \frac{y^2}{12} = 1$$

which is bisected at the point (4, 2).

21. Required the equations of the tangents to

$$\frac{x^2}{16} - \frac{y^2}{10} = 1$$

which make angles of 60° with the X-axis.

22. Show that the rectangle of the distances intercepted on the tangents drawn at the vertices of an hyperbola by a tangent drawn at any point is constant and equal to the square of the semi-conjugate axis.

23. Given the base of a triangle and the difference of the tangents of the base angles; required the locus of the vertex.

24. Show that the polars of (m, n) with respect to the hyperbolas

$$\frac{x^2}{a^2} - \frac{y^2}{b^2} = 1, \quad \frac{y^2}{b^2} - \frac{x^2}{a^2} = 1 \text{ are parallel.}$$

25. If from the foot of the ordinate of a point (x, y) of the hyperbola a tangent be drawn to the circle constructed on the transverse axis, and from the point of tangency a line be drawn to the centre, the angle which this line forms with the transverse axis is called the eccentric angle of (x, y). Show that (x, y) = (a sec φ, b tan φ), and from these values deduce the equation of the hyperbola.

26. If (x', y'), (x'', y'') are the extremities of a pair of conjugate diameters whose eccentric angles are φ' and φ, show that $\varphi' + \varphi = 90°$.

CHAPTER IX.

THE GENERAL EQUATION OF THE SECOND DEGREE.

136. The most general equation of the second degree between two variables is

$$ay^2 + bxy + cx^2 + dy + ex + f = 0 \ldots (1)$$

in which a, b, c, d, e, f are any constant quantities whatever. To investigate the properties of the loci which this equation represents under all possible values of the constants as to sign and magnitude is the object of this chapter.

137. The equations of the lines in a plane, with which we have had to do in preceding chapters, are

$Ax + By + C = 0$. Straight line.

$(Ax + By + C)^2 = 0$. Two coincident straight lines.

$y^2 - x^2 = 0$. Two straight lines.

$y^2 + x^2 = a^2$. Circle.

$y^2 + x^2 = 0$. Two imaginary straight lines.

$y^2 = 2\,px$. Parabola.

$a^2y^2 + b^2x^2 = a^2b^2$. Ellipse.

$a^2y^2 - b^2x^2 = -a^2b^2$. Hyperbola.

$a^2y^2 - b^2x^2 = a^2b^2$. Hyperbola.

Comparing these equations with the general equation, we see that all of them may be deduced from it by making the constants fulfil certain conditions as to sign and magnitude. We are, therefore, prepared to expect that the lines which these equations represent will appear among the loci represented by the general equation of the second degree between two variables. In the discussion which is to ensue we shall find that these lines are the *only* loci represented by this equation.

DISCUSSION.

138. *To show that the locus represented by a complete equation of the second degree between two variables is also represented by an equation of the second degree between two variables, in which the term containing xy is wanting.*

Let us assume the equation

$$ay^2 + bxy + cx^2 + dy + ex + f = 0 \ \cdots \ (1)$$

and refer the locus it represents to rectangular axes, making the angle θ with the old axes, the origin remaining the same. From Art. 33, Cor. 2, we have

$$x = x' \cos \theta - y' \sin \theta$$
$$y = x' \sin \theta + y' \cos \theta$$

for the equations of transformation. Substituting these values in (1), we have,

$$a'y'^2 + b'x'y' + c'x'^2 + d'y + e'x' + f = 0 \ \cdots \ (2)$$

in which

$$
\left.
\begin{aligned}
a' &= a \cos^2 \theta + c \sin^2 \theta - b \sin \theta \cos \theta \\
b' &= 2(a-c) \sin \theta \cos \theta + b(\cos^2 \theta - \sin^2 \theta) \\
c' &= a \sin^2 \theta + c \cos^2 \theta + b \sin \theta \cos \theta \\
d' &= d \cos \theta - e \sin \theta \\
e' &= d \sin \theta + e \cos \theta
\end{aligned}
\right\} \ \cdots \ (3)
$$

Since θ, the angle through which the axes have been turned, is entirely arbitrary, we are at liberty to give it such a value as will render the value of b' equal to zero. Supposing it to have that value, we have

$$2(a - c) \sin \theta \cos \theta + b(\cos^2 \theta - \sin^2 \theta) = 0,$$

or $\quad (a - c) \sin 2\theta + b \cos 2\theta = 0 \ \cdots \ (4)$

or $\quad\quad\quad \tan 2\theta = \dfrac{b}{c-a} \ \cdots \ (5)$

Since any real number between $+\infty$ and $-\infty$ is the tangent of some angle, equation (5) will always give real value for 2θ; hence the above transformation is always possible. Making $b' = 0$ in (2), we have, dropping accents,

$$a'y^2 + c'x^2 + d'y + e'x + f = 0 \ \cdots \ (6)$$

for the equation of the locus represented by (1). To this equation, then, we shall confine our attention.

172 PLANE ANALYTIC GEOMETRY.

Cor. 1. *To find the value of a' and c' in terms of a, b, and c.* Adding and then subtracting the *first* and *third* of the equations in (3), we have

$$c' + a' = c + a \ \ldots (7)$$
$$c' - a' = (c - a)\cos 2\theta + b \sin 2\theta \ \ldots (8)$$

Squaring (4) and adding to the square of (8), we have

$$(c' - a')^2 = (c - a)^2 + b^2;$$
$$\therefore c' - a' = \sqrt{(c - a)^2 + b^2} \ \ldots (9)$$

Subtracting and then adding (7) and (9), we have

$$a' = \tfrac{1}{2}\{c + a - \sqrt{(c - a)^2 + b^2}\} \ \ldots (10)$$
$$c' = \tfrac{1}{2}\{c + a + \sqrt{(c - a)^2 + b^2}\} \ \ldots (11)$$

Cor. 2. *To find the signs of a' and c'.* Multiplying (10) and (11), we have

$$a'c' = \tfrac{1}{4}\{(c + a)^2 - ((c - a)^2 + b^2)\};$$
$$\therefore a'c' = -\tfrac{1}{4}(b^2 - 4ac) \ \ldots (12)$$

Hence, the *signs of a' and c' depend upon the sign of the quantity $b^2 - 4ac$.*

The following cases present themselves:

1. $b^2 < 4ac$. The sign of the second member of (12) is positive, $\therefore a'$ and c' are both *positive*, or both *negative*.

2. $b^2 = 4ac$. The second member of (12) becomes zero, $\therefore a' = 0$, or $c' = 0$.

[It will be observed that a' and c' cannot be equal to zero at the same time, for such a supposition would reduce (6) to an equation of the first degree.]

3. $b^2 > 4ac$. The sign of the second member of (12) is negative, $\therefore a'$ must be *positive* and c' *negative*, or a' must be *negative* and c' *positive*.

139. *To transform the equation $a'y^2 + c'x^2 + d'y + e'x + f = 0$ into an equation in which the first powers of the variables are missing.*

Let us refer the locus to a parallel system of rectangular axes, the origin being at the point (m, n). From Art. 32, we have

$$x = m + x', \; y = n + y'.$$

Substituting these values in the given equation, we have

$$a'y'^2 + c'x'^2 + d''y' + e''x' + f'' = 0 \; \cdots \; (2)$$

in which

$$\left.\begin{array}{l} d'' = 2\,a'n + d' \\ e'' = 2\,c'm + e' \\ f'' = a'n^2 + c'm^2 + d'n + e'm + f \end{array}\right\} \; \cdots \; (3)$$

Since m and n are entirely arbitrary, we may, *in general,* give them such values as to make

$$2\,a'n + d' = 0 \text{ and } 2\,c'm + e' = 0;$$

i.e., in general, we may make

$$n = -\frac{d'}{2\,a'} \text{ and } m = -\frac{e'}{2\,c'} \; \cdots \; (4)$$

We see from these values that when a' and c' are *not zero,* this transformation also is possible ; and equation (2) becomes, after dropping accents,

$$a'y^2 + c'x^2 + f'' = 0 \; \cdots \; (5)$$

Equation (5), we observe, contains only the second power of the variables; hence it is satisfied for the points (x, y) and $(-x, -y)$. But only the equation of curves with centres can satisfy this condition; hence, equation (5) is the equation of *central loci.* When either a' or c' is zero, then n or m is *infinite* and the transformation becomes impossible. Hence arise two cases which require special consideration.

140. CASE 1. $a' = o.$

Under this supposition equation (6), Art. 138, becomes

$$c'x^2 + d'y + e'x + f = 0 \; \cdots \; (1)$$

Referring the locus of this equation to parallel axes, the origin being changed, we have for the equations of transformation

$$x = m + x', \; y = n + y'.$$

Substituting in (1), we have

$$c'x'^2 + d'y' + (2\,c'm + e')\,x' + c'm^2 + d'n + e'm + f = 0 \ldots (2)$$

Now, *in general,* we may give m and n such values as to make

$$2\,c'm + e' = 0, \text{ and } c'm^2 + d'n + e'm + f = 0\,;$$

i.e., we may make

$$\left.\begin{aligned} m &= -\frac{e'}{2\,c'}, \text{ and} \\ n &= -\frac{c'm^2 + e'm + f}{d'} = \frac{e'^2 - 4\,fc'}{4\,d'c'} \end{aligned}\right\} \ \cdots \ (a)$$

If d' is not zero (since $a' = 0$, c' is not zero), this transformation is possible and (2) becomes, after dropping accents,

$$c'x^2 + d'y = 0,$$

or

$$x^2 = -\frac{d'}{c'}\,y \ \ldots \ (3)$$

Cor. If $d' = 0$, (1) becomes

$$c'x^2 + e'x + f = 0 \ \ldots \ (4)$$

or, solving with respect to x,

$$x = \frac{-\,e' \pm \sqrt{e'^2 - 4\,fc'}}{2\,c'} \ \ldots \ (5)$$

141. Case 2. $c' = o.$

Under this supposition equation (6), Art. 138, becomes

$$a'y^2 + d'y + e'x + f = 0 \ \ldots \ (1)$$

Transforming this equation so as to eliminate y and the constant term, by a method exactly similar to that of the preceding article, we find

$$n = -\frac{d'}{2\,a'},$$

$$m = \frac{d'^2 - 4\,a'f}{4\,a'e'}\,;$$

and, if e' is not zero, we have (a' is not zero since $c' = 0$)

$$y^2 = -\frac{e'}{a'}\,x \ \ldots \ (2)$$

Cor. If $e' = 0$, equation (1) becomes

$$a'y^2 + d'y + f = 0,$$

or

$$y = \frac{-d' \pm \sqrt{d'^2 - 4fa'}}{2\,a'} \quad \ldots \quad (3)$$

142. Summarizing the results of the preceding articles, we find that the discussion of the general equation

$$ay^2 + bxy + cx^2 + dy + ex + f = 0$$

has been reduced to the discussion of the three simple forms:

1. $a'y^2 + c'x^2 + f'' = 0$.   Art. 139, (5)

2. $\begin{cases} x^2 = -\dfrac{d'}{c'}\, y. & \text{Art. 140, (3)} \\[2mm] y^2 = -\dfrac{e'}{a'}\, x. & \text{Art. 141, (2)} \end{cases}$

3. $\begin{cases} x = \dfrac{-e' \pm \sqrt{e'^2 - 4fc'}}{2\,c'}. & \text{Art. 140, (5)} \\[2mm] y = \dfrac{-d' \pm \sqrt{d'^2 - 4fa'}}{2\,a'}. & \text{Art. 141, (3)} \end{cases}$

The discussion now involves merely a consideration of the sign and magnitude of the constants which enter into these equations.

143. $b^2 < 4\,ac$.

Under this supposition, since a' and c' are both *positive* or both *negative*, Art. 138, Cor. 2, neither a' nor c' can be zero; hence, forms 2 and 3 of the preceding article are excluded from consideration.

The first form becomes either

$$\left. \begin{array}{l} a'y^2 + c'x^2 + f'' = 0, \\ -a'y^2 - c'x^2 + f'' = 0 \end{array} \right\} \quad \ldots \quad (1)$$

or

in which a' and c' may have any real value and f'' may have any sign and any value. Hence arise four cases:

CASE 1. If f'' has a sign different from that of a' and c', equations (1) are *equations of ellipses* whose semi-axes are

$$a = \sqrt{\frac{f''}{c'}} \text{ and } b = \sqrt{\frac{f''}{a'}}.$$

CASE 2. If f'' has the same sign as that of a' and c', equations (1) represent *imaginary curves*.

CASE 3. If $a' = c'$ and f'' has a different sign from that of a' and c', equations (1) are *equations of circles*. If f'' has the same sign as a' and c', then the equations represent *imaginary curves*.

CASE 4. If $f'' = 0$, equations (1) are equations of *two imaginary straight lines passing through the origin.*

Hence, when $b^2 < 4\,ac$, *every equation of the second degree between two variables represents an ellipse, an imaginary curve, a circle, or two imaginary straight lines intersecting at the origin.*

144. $b^2 = 4\,ac.$

Under this supposition, Art. 138, Cor. 2, either $a' = 0$, or $c' = 0$; hence, form (1) of Art. 142 is excluded.

Resuming the forms

$$\left. \begin{aligned} x^2 &= -\frac{d'}{c'}\, y \\[2mm] y^2 &= -\frac{e'}{a'}\, x \end{aligned} \right\} \quad \dots (2)$$

$$\left. \begin{aligned} x &= \frac{-e' \pm \sqrt{e'^2 - 4\,fc'}}{2\,c'} \\[2mm] y &= \frac{-d' \pm \sqrt{d'^2 - 4\,fa'}}{2\,a'} \end{aligned} \right\} \quad \dots (3)$$

we have four cases depending upon the sign and magnitude of the constants.

CASE 1. If d' and c' in the first form of (2) are not zero, and if e' and a' in the second form of (2) are not zero, then equations (2) are equations of *parabolas*.

CASE 2. Since the first form of (3) is independent of y, it represents *two lines* parallel to each other and to the Y-axis. The second form of (3) represents, similarly, *two lines* which are parallel to the X-axis.

CASE 3. If $e'^2 < 4fc'$ the first form of (3) represents *two imaginary lines.*

If $d'^2 < 4fa'$, the second form of (3) represents *two imaginary lines.*

CASE 4. If $e'^2 = 4fc'$, the first form of (3) represents *one straight line* parallel to the Y-axis.

If $d'^2 = 4fa'$, the second form of (3) represents *one straight line* parallel to the X-axis.

Hence, *when $b^2 = 4ac$, every equation of the second degree between two variables represents a parabola, two parallel straight lines, two imaginary lines, or one straight line.*

145. $b^2 > 4ac.$

Under this supposition, Art. 138, Cor. 2, since a' and c' must have opposite signs, neither a' nor c' can be zero; hence forms (2) and (3) of Art. 142 are excluded from consideration under this head. The first form becomes either

$$a'y^2 - c'x^2 + f'' = 0 \atop -a'y^2 + c'x^2 + f'' = 0 \Big\} \; \cdots \; (1)$$

or

We have here three cases.

CASE 1. If f'' has a different sign from that of a', equations (1) are *equations of hyperbolas* whose semi-axes are

$$a = \sqrt{\frac{f''}{c'}} \text{ and } b = \sqrt{\frac{f''}{a'}}.$$

If f'' has a different sign from that of c', equations (1) are still *equations of hyperbolas.*

CASE 2. If $a' = c'$, equations (1) are *equations of equilateral hyperbolas.*

CASE 3. If $f'' = 0$, equations (1) are *equations of two intersecting straight lines.*

Hence, *when $b^2 > 4\,ac$, every equation of the second degree between two variables represents an hyperbola, an equilateral hyperbola, or two intersecting straight lines.*

146. SUMMARY. The preceding discussion has elicited the following facts :

1. *That the general equation of the second degree between two variables represents, under every conceivable value of the constants which enter into it, an ellipse, a parabola, an hyperbola, or one of their limiting cases.*

2. *When $b^2 < 4ac$ it represents an ellipse, or a limiting case.*

3. *When $b^2 = 4ac$ it represents a parabola, or a limiting case.*

4. *When $b^2 > 4ac$ it represents an hyperbola, or a limiting case.*

EXAMPLES.

1. Given the equation $3\,y^2 + 2\,xy + 3\,x^2 - 8\,y - 8\,x = 0$; to classify the locus, transform and construct the equation.

(*a*) *To classify.* Write the general equation and just below it the given equation, thus :

$$ay^2 + bxy + cx^2 + dy + ex + f = 0$$
$$3\,y^2 + 2\,xy + 3\,x^2 - 8\,y - 8\,x = 0 \ \ldots \ (1)$$

Substituting the co-efficients in the class characteristic $b^2 - 4\,ac$, we have $\quad b^2 - 4\,ac = 4 - 36 = -32$;
hence $\qquad b^2 < 4\,ac.$
and the locus belongs to the ellipse class, Art. 146.

(*b*) *To refer the locus to axes such that the term containing xy shall disappear.*

From Art. 138, (5), we have

$$\tan 2\,\theta = \frac{b}{c - a};$$

hence $\qquad \tan 2\,\theta = \dfrac{2}{3 - 3} = +\infty,$

$$\therefore 2\,\theta = 90° \ \therefore \ \theta = 45° \ \ldots \ (2)$$

i.e., the new X-axis makes an angle of $+45°$ with the old X-axis. Taking now (10), (11), (3), Art. 138, and substituting values, we have

$$a' = \tfrac{1}{2}\{c + a - \sqrt{(c - a)^2 + b^2}\} = 2.$$
$$c' = \tfrac{1}{2}\{c + a + \sqrt{(c - a)^2 + b^2}\} = 4.$$
$$d' = d \cos\theta - e \sin\theta = \tfrac{1}{2}\sqrt{2}\,(d - e) = 0.$$
$$e' = d \sin\theta + e \cos\theta = \tfrac{1}{2}\sqrt{2}\,(d + e) = -8\sqrt{2}.$$

Substituting these values in (6), Art. 138, we have $(f$ being zero),

$$2\,y^2 + 4\,x^2 - 8\sqrt{2}\,.\,x = 0 \dots (3)$$

(c) *To refer the locus to its centre and axes.*
Substituting the values found above in (4), Art. 139, we

have $\qquad n = -\dfrac{d'}{2\,a'} = 0.$

$$m = -\frac{e'}{2\,c'} = \frac{8\sqrt{2}}{8} = \sqrt{2}.$$

Hence $f'' = a'n^2 + c'm^2 + d'n + e'm + f = -8$, Art. 139, (3).
Substituting this value of f'' together with the values of a' and c' found above in (5), Art. 139, we have

$$2\,y^2 + 4\,x^2 - 8 = 0,$$

or $\qquad \dfrac{x^2}{2} + \dfrac{y^2}{4} = 1 \dots (4)$

for the reduced equation. The semi-axes of the ellipse are $a = \sqrt{2}$ and $b = 2$.

(d) *To construct.*

Draw the axis OX', making an angle of 45° with the old X-axis. See (b). Draw OY' $\perp$ to OX'. The equation of the curve when referred to these axes is given in (3). Constructing the point O' ($\sqrt{2}$, 0) we have the centre of the ellipse. See (*c*). Draw $O'Y'' \perp$ to OX' at O'. The equation of the curve when referred to $O'Y''$, $O'X'$ as axes is given in (4).

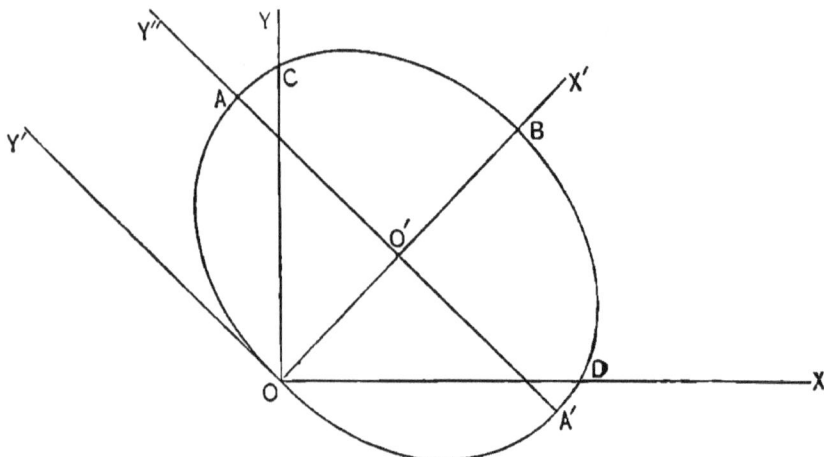

FIG. 55.

Having the semi-axes, $\sqrt{2}$ and 2, we can construct the ellipse by either of the methods given in Art. 78.

DISCUSSION.

If $y = 0$ in (1), we have for the X-intercepts O, OD,

$$x = 0, \ x = \frac{8}{3}.$$

If $x = 0$ in (1), we have for the Y-intercepts O, OC,

$$y = 0, \ y = \frac{8}{3}.$$

If $x = 0$ in (3), we have $y = \pm 0$; i.e., the ellipse is tangent to the Y'-axis.

If $y = 0$ in (3), we have for the X'-intercepts O, OB,

$$x = 0, \ x = 2\sqrt{2}.$$

If $x = 0$ in (4), we have for the Y''-intercepts O'A, O'A'

$$y = \pm 2.$$

If $y = 0$ in (4), we have for the X'-intercepts O'B, O'O,

$$x = \pm \sqrt{2}.$$

2. Given the equation $y^2 - 2\,xy + x^2 - 2\,y - 1 = 0$, classify the locus, transform and construct the equation.

(a) *To classify.*

$$ay^2 + bxy + cx^2 + dy + ex + f = 0$$
$$y^2 - 2\,xy + x^2 - 2\,y - 1 = 0 \;\ldots\; (1)$$

hence
$$b^2 - 4\,ac = 4 - 4 = 0,$$
$$\therefore b^2 = 4\,ac\,;$$

hence the locus belongs to the parabola class, Art. 146.

(b) *To refer the locus to axes such that the term containing xy shall disappear.*

From Art. 138, (5), we have

$$\tan 2\,\theta = \frac{b}{c - a}\,;$$

hence, substituting

$$\tan 2\,\theta = - \frac{2}{1 - 1} = - \infty;$$
$$\therefore \theta = - 45^\circ \;\ldots\; (2)$$

Substituting the values of the coefficients in (10), (11), (3) of Art. 138, we have

$$a' = \tfrac{1}{2}\{c + a - \sqrt{(c - a)^2 + b^2}\} = 0.$$
$$c' = \tfrac{1}{2}\{c + a + \sqrt{(c - a)^2 + b^2}\} = 2.$$
$$d' = d \cos \theta - e \sin \theta = - 2 \,(\tfrac{1}{2} \sqrt{2}) = - \sqrt{2}.$$
$$e' = d \sin \theta + e \cos \theta = - 2 \,(- \tfrac{1}{2} \sqrt{2}) = + \sqrt{2}.$$

Substituting these values in (1), Art. 140 (since $a' = 0$), we have
$$2\,x^2 - \sqrt{2}\,y + \sqrt{2}\,x - 1 = 0 \;\ldots\; (3)$$

(c) *To refer the parabola to a tangent at the vertex and the axis.*

Substituting the values of the constants in (a), Art. 140, we have
$$m = - \frac{e'}{2\,c'} = - \frac{\sqrt{2}}{4} = - .35 \text{ nearly.}$$

$$n = \frac{e'^2 - 4\,fc'}{4\,d'c'} = - \frac{5}{4\sqrt{2}} = - .90 \text{ nearly.}$$

Substituting the values of d' and c' in (3), Art. 140 (since d' is not zero), we have

$$x^2 = \tfrac{1}{2} \sqrt{2} \cdot y \ \ldots \ (4)$$

for the reduced equation.

(d) *To construct.*

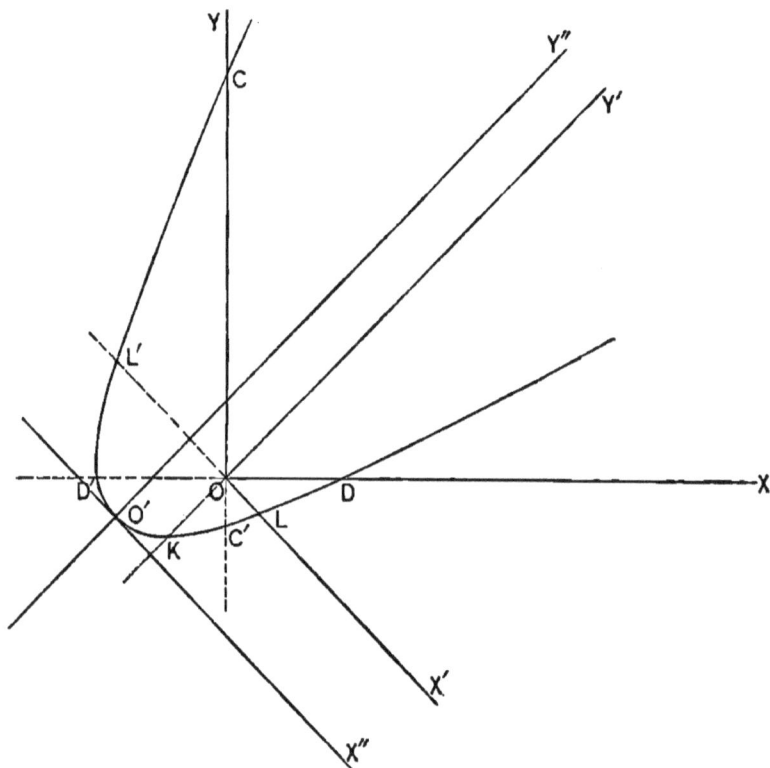

FIG. 56.

Draw OX' making an angle of $-45°$ with the X-axis; draw $OY' \perp$ to OX'. See (b). The equation of the parabola when referred to these axes is given in (3).

Constructing the point $(-.35, -.90)$, we have the vertex of the parabola O'. See (c). Draw $O'X''$ and $O'Y''$ parallel to the axes OX', OY' respectively. The equation of the parabola referred to these axes is given in (4). The curve can now be constructed by either of the methods given in Art. 54.

DISCUSSION.

If $x = 0$ in (1), we have for the Y-intercepts OC, OC′,

$$y = 2.4 \qquad y = -.4.$$

If $y = 0$ in (1), we have for the Y-intercept OD, OD′,

$$x = \pm 1.$$

If $x = 0$ in (3), we have for the Y′-intercept OK,

$$y = -\frac{1}{\sqrt{2}} = -.707.$$

If $y = 0$ in (3), we have for the X′-intercepts OL, OL′,

$$x = \frac{-\sqrt{2} + \sqrt{10}}{4}, \qquad x = \frac{-\sqrt{2} - \sqrt{10}}{4}.$$

If $x = 0$ in (4), $y = 0$; if $y = 0$ in (4), $x = \pm 0$.

3. Given the equation $y^2 - 2x^2 - 2y + 6x - 3 = 0$, classify the locus, transform and construct the equation.

(a) *To classify.*

$$ay^2 + bxy + cx^2 + dy + ex + f = 0.$$
$$y^2 - 2x^2 - 2y + 6x - 3 = 0 \;\ldots\; (1)$$
$$b^2 - 4ac = 8 \;\therefore\; b^2 > 4ac;$$

hence, the locus belongs to the hyperbola class, Art. 146.

(b) To ascertain the direction of the rectangular axes (xy being wanting).

$$\tan 2\theta = \frac{b}{c - a} = \frac{0}{-3} = 0;$$
$$\therefore\; \theta = 0;$$

i.e., the new X-axis is parallel to the old X-axis.

(c) *To refer the hyperbola to its centre and axes*, we have, Art. 139, (4),

$$n = -\frac{d'}{2\,a'}, \; m = -\frac{e'}{2\,c'};$$

hence $\quad n = 1, \, m = \dfrac{3}{2}.$

Substituting in the value of f'', Art. 139, (3), we have

$$f'' = a'n^2 + c'm^2 + d'n + e'm + f = 1 - \frac{9}{2} - 2 + 9 - 3;$$

hence $\quad f'' = \dfrac{1}{2}.$

This value, together with the values of a' and c' in (5), Art. 139, gives $2\,y^2 - 4\,x^2 = -1\ \ldots$ (3) for the required equation.

(*d*) *To construct.*

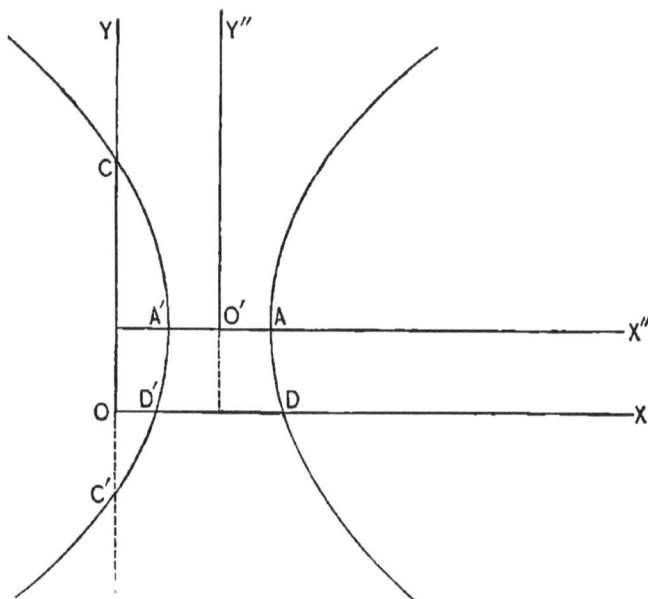

FIG. 57.

Construct the point O' $(\frac{3}{2}, 1)$, and through it draw O'X″ ∥ to OX, and O'Y″ ∥ to OY. The equation of the hyperbola referred to these axes is given in (3). We see from this equation that the semi-transverse axis is $\frac{1}{2}$. Laying off this distance to the right and then to the left of O', we locate the vertices of the curve A, A'.

DISCUSSION.

If $x = 0$ in (1), we have for the Y-intercepts OC, OC',

$$y = 3,\ y = -1.$$

If $y = 0$ in (1), we have for the X-intercepts OD, OD',

$$x = \frac{3 + \sqrt{3}}{2}, \qquad x = \frac{3 - \sqrt{3}}{2}.$$

If $x = 0$ in (3), we have

$$y = \pm \sqrt{-\tfrac{1}{3}}.$$

If $y = 0$ in (3), we have for the X-intercepts O'A, O'A',

$$x = \pm \frac{1}{2}.$$

From this data the student may readily determine the eccentricity, the parameter, and the focal distances of the hyperbola.

4. Given the equation $y^2 + x^2 - 4y + 4x - 1 = 0$, classify the locus, transform and construct the equation.

(a) $b^2 < 4\,ac$ ∴ the locus belongs to the ellipse class.

(b) $\theta = 0$ ∴ new X-axis is ∥ to old X-axis.

(c) $(m, n) = (-2, 2)$ and $f'' = -9$
hence $x^2 + y^2 = 9$

is the transformed equation of the locus, which from the form of the equation is evidently a circle.

(d) Locate the point $(-2, 2)$. With this point as a centre, and with 3 as a radius, describe a circle; it will be the required locus.

5. $y^2 - 2\,xy + x^2 - 2 = 0$.

(a) $b^2 = 4\,ac$ ∴ parabola class.

(b) $\theta = -45°$ ∴ new X-axis inclined at an angle of $-45°$ to the old X-axis. We have also

$$a' = 0,\ c' = 2,\ d' = 0,\ e' = 0$$
$$\therefore 2\,x^2 - 2 = 0 ;$$
$$\text{i.e., } x = 1 \text{ and } x = -1 \ . \ . \ . \ (1)$$

are the equations of the locus when referred to the new axes.

(c) The construction gives the lines OX', OY' as the new axes of reference.

Equations (1) are the equations of the two lines CM, C'M' drawn ∥ to the Y'-axis and at a unit's distance from it.

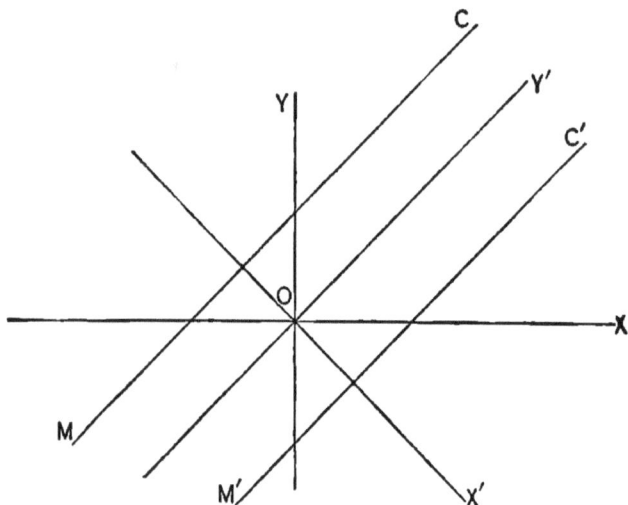

We may construct the locus of the given equation without going through the various steps required by the general method. Factoring the given equation, we have

$$(y - x + \sqrt{2})\,(y - x - \sqrt{2}) = 0\,;$$

hence $\qquad y = x - \sqrt{2}$ and $y = x + \sqrt{2}$

are the equations of the locus. Constructing these lines (OY, OX being the axes of reference), we get the two parallel lines CM, C′M′.

Classify, transform, and construct each of the following equations:

6. $y^2 - 2xy + x^2 + 2y - 2x + 1 = 0.$

$$x = \tfrac{1}{2}\sqrt{2}$$

7. $y^2 + 2xy + x^2 - 1 = 0.$

$$x = \pm \tfrac{1}{2}\sqrt{2}$$

8. $5y^2 + 2xy + 5x^2 - 12x - 12y = 0.$

$$\frac{x^2}{2} + \frac{y^2}{3} = 1.$$

9. $2\,y^2 + 2\,x^2 - 4\,y - 4\,x + 1 = 0.$

$x^2 + y^2 = \tfrac{3}{2}.$

10. $y^2 + x^2 - 2\,x + 1 = 0.$

$y = x\,\sqrt{-1},\ (0,\,0).$

11. $y^2 + x^2 + 2\,x + 2 = 0.$

Imaginary ellipse.

12. $y^2 - 2\,xy + x^2 - 8\,x + 16 = 0.$

Parabola.

13. $y^2 - 2\,xy + x^2 - y + 2\,x - 1 = 0.$

Parabola.

14. $4\,xy - 2\,x + 2 = 0.$

Hyperbola.

15. $y^2 - 2\,x^2 + 2\,y + 1 = 0.$

Two intersecting lines.

16. $y^2 - x^2 + 2\,y + 2\,x - 4 = 0.$

Equilateral hyperbola.

17. $y^2 - 2\,xy + x^2 + 2\,y + 1 = 0.$
18. $y^2 + 4\,xy + 4\,x^2 - 4 = 0.$
19. $y^2 - 2\,xy + 2\,x^2 - 2\,y + 2\,x = 0.$
20. $y^2 - 4\,xy + 4\,x^2 = 0.$
21. $y^2 - 2\,xy - x^2 + 2 = 0.$
22. $y^2 - x^2 = 0.$

CHAPTER X.

HIGHER PLANE CURVES.

147. Loci lying in a single plain and represented by equations other than those of the first and second degrees are called HIGHER PLANE CURVES. We shall confine our attention in this chapter to the consideration of a few of those curves which have become celebrated by reason of the labor expended upon them by the ancient mathematicians, or which have become important by reason of their practical value in the arts and sciences.

EQUATIONS OF THE THIRD DEGREE.

148. THE SEMI-CUBIC PARABOLA.

This curve is the locus generated by the intersection of the ordinate TT' of the common parabola with the perpendicular OP let fall from its vertex upon the tangent drawn at T' as the point of tangency moves around the curve.

1. *To deduce the rectangular equation.*

Let T' (x'', y'') be the point of tangency, and let P (x, y) be a point of the curve.

Let $y^2 = 4\,px$ be the equation of the common parabola.

Since the equation of the tangent line T'M to the parabola is Art. 57, (6),

$$yy'' = 2\,p(x + x''),$$

the equation of the perpendicular (OM) let fall from the vertex is

$$y = -\frac{y''}{2\,p}\,x \ \ldots \ (1)$$

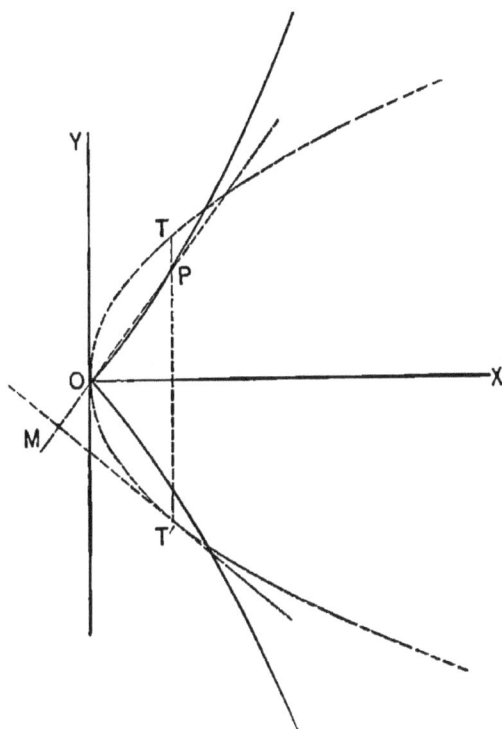

FIG. 59.

Since TT' is parallel to OY, we have for its equation

$$x = x'' \ldots (2)$$

Combining (1) and (2), we have

$$y = - \frac{y''}{2p} x''.$$

But $\qquad y'' = \sqrt{4px''} \,;$

hence $\qquad y = - \frac{\sqrt{4px''}}{2p} x''.$

Squaring and dropping accents, we have

$$y^2 = \frac{x^3}{p} \ldots (3)$$

for *the equation of the semi-cubic parabola.*

This curve is remarkable as being the first curve which was *rectified*, that is, the length of a portion of it was shown to

be equal to a certain number of rectilinear units. It derives
its name from the fact that its equation (3) may be written

$$x^{\frac{3}{2}} = p^{\frac{1}{2}} y.$$

2. *To deduce the polar equation.*

Making $x = r \cos \theta$ and $y = r \sin \theta$ in (3), we have, after
reduction,

$$r = p \tan^2 \theta \sec \theta \ \ldots \ (4)$$

for the polar equation of the curve.

Schol. Solving (3) with respect to y, we have

$$y = \pm \sqrt{\frac{x^3}{p}}.$$

An inspection of this value shows

(*a*) That the curve is symmetrical with respect to the
X-axis;

(*b*) That the curve extends infinitely from the Y-axis in
the direction of the positive abscissas.

149. *To duplicate the cube by the aid of the parabola.*

Let a be the edge of the given cube. We wish to *construct* the edge of a cube such that the cube constructed on it
shall be double the volume of the given cube; i.e., that the
condition $x^3 = 2 a^3$ shall be satisfied.

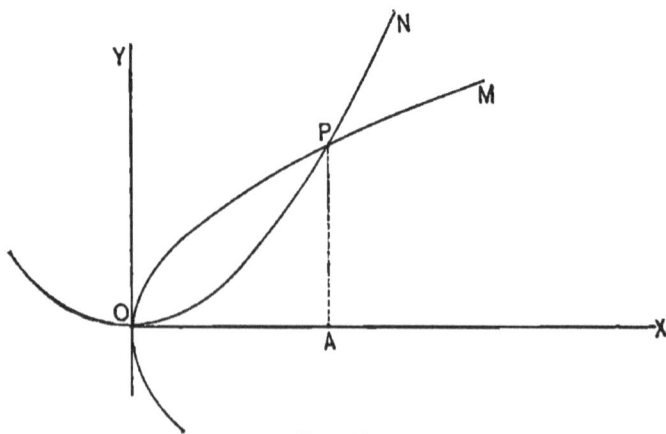

FIG. 60.

Construct the parabola whose equation is

$$y^2 = 2ax \ldots (1)$$

Let MPO be the curve. Construct also the parabola whose equation is

$$x^2 = ay \ldots (2)$$

Let NPO be this curve.

Then OA ($= x$), the abscissa of their point of intersection is the required edge. For eliminating y between (1) and (2), we have

$$x^3 = 2a^3.$$

This problem attained to great celebrity among the ancient geometricians. We shall point out as we proceed one of the methods employed by them in solving it.

150. The Cissoid.

The cissoid is the locus generated by the intersection (P) of the chord (OM') of the circle (OMM'T) with the ordinate

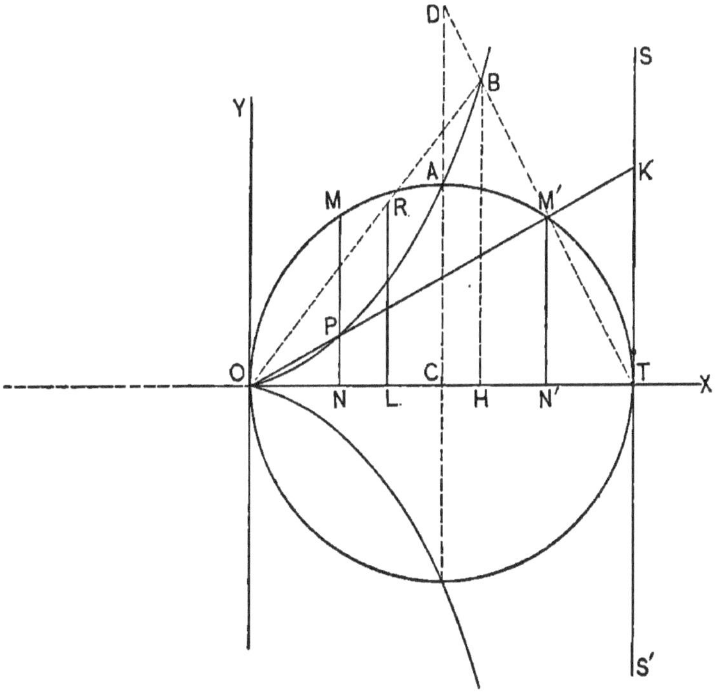

Fig. 61.

MN (equal to the ordinate M'N' let fall from the point M' on the diameter through O) as the chord revolves about the origin O.

It may also be defined as the locus generated by the intersection of a tangent to the parabola $y^2 = -8ax$ with the perpendicular let fall on it from the origin as the point of tangency moves around the curve.

1. *To deduce the rectangular equation.*

First Method. — Let $OT = 2a$, and let P (x, y) be any point of the curve. From the method of generation in this case $MN = M'N' \therefore ON = N'T$. From the similar triangles ONP, ON'M', we have

$$NP : ON :: M'N' : ON'.$$

But $NP = y$, $ON = x$, $M'N' = \sqrt{ON' . N'T} = \sqrt{(2a-x)x}$, $ON' = 2a - x$;

$$\therefore y : x :: \sqrt{(2a-x)x} : 2a - x.$$

Hence $\qquad y^2 = \dfrac{x^3}{2a-x} \quad \cdots \quad (1)$

is the required equation.

Second Method. — The equation of the tangent line to the parabola $y^2 = -8ax$ is Art. 65, (2)

$$y = -sx + \frac{2a}{s}.$$

The equation of a line passing through the origin and perpendicular to this line is

$$y = \frac{x}{s}.$$

Combining these equations so as to eliminate s, we have

$$y^2 = \frac{x^3}{2a-x}$$

for the equation of the locus.

This curve was invented by Diocles, a Greek mathematician of the second century, B.C., and called by him the cissoid from a Greek word meaning "ivy." It was employed by him in solving the celebrated problem of inserting two mean propor- tionals between given extremes, of which the duplication of the cube is a particular case.

2. *To deduce the polar equation.*

From the figure $(OP, PON) = (r, \theta)$
we have also $r = OP = M'K = OK - OM'$.

But $OK = 2\,a \sec \theta$ and $OM' = 2\,a \cos \theta$; hence

$$r = 2\,a\ (\sec \theta - \cos \theta),$$
or $\qquad\qquad r = 2\,a \tan \theta \sin \theta$

is the polar equation of the curve.

SCHOL. Solving (1) with respect to y, we have

$$y = \pm \sqrt{\frac{x^3}{2\,a - x}}.$$

An inspection of this value shows

(*a*) That the cissoid is symmetrical with respect to the X-axis.

(*b*) That $x = 0$ and $x = 2\,a$ are the equations of its limits.

(*c*) That $x = 2\,a$ is the equation of a rectilinear asymp- tote (SS').

151. *To duplicate the cube by the aid of the cissoid.*

Let OL, Fig. 61, be the edge of the cube which we wish to duplicate. Construct the arc BO of the cissoid, $CO = a$ being the radius of the base circle. Lay off $CD = 2\,CA = 2\,a$ and draw DT intersecting the cissoid in B; draw BO and at L erect the perpendicular LR intersecting BO in R. Then LR is the edge of the required cube; for the equation of the cissoid gives

$$y^2 = \frac{x^3}{2\,a - x};$$

hence $\qquad HB^2 = \dfrac{OH^3}{HT}$ (since $HB = y$, $OH = x$, and $HT = 2\,a - x$).

The similar triangles CDT and HBT give
$$CD : CT :: HB : HT.$$

But $CD = 2\,CT$ by construction; hence $HB = 2\,HT$

$$\therefore HT = \frac{HB}{2}.$$

This value of HT in the value of HB^2 above gives

$$HB^2 = \frac{2\,OH^3}{HB}; \text{ hence } HB^3 = 2\,OH^3.$$

The triangles OHB and OLR are similar; hence

$$HB : OH :: LR : OL$$
$$\therefore HB^3 : OH^3 :: LR^3 :: OL^3$$

But $HB^3 = 2\,OH^3$, hence $LR^3 = 2\,OL^3$; whence the construction.

152. THE WITCH.

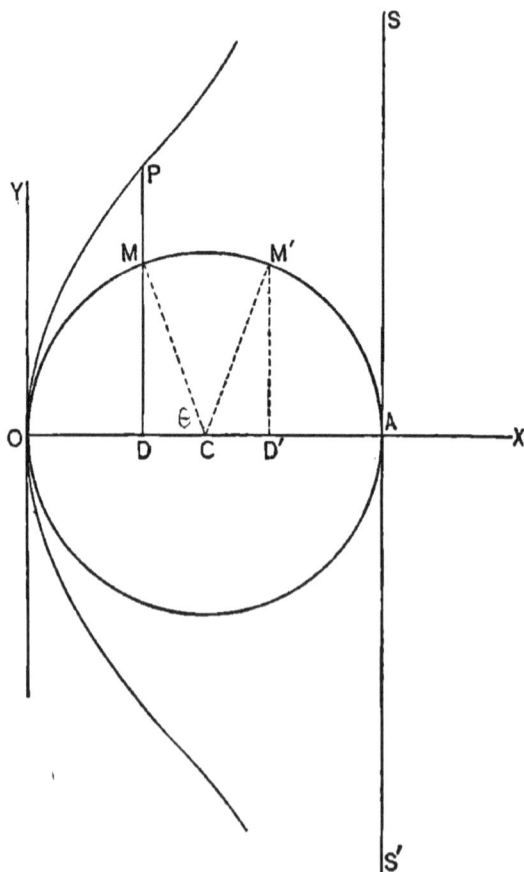

FIG. 62.

The witch is the locus of a point P on the produced ordinate DP of a circle, so that the produced ordinate DP is to the diameter of the circle OA as the ordinate DM is to the outer segment DA of the diameter.

It may also be defined as the locus of a point P on the linear sine DM of an angle at a distance from its foot D equal to twice the linear tangent of one-half the angle.

1. *To deduce the rectangular equation.*

First Method. — From the mode of generation, we have

$$DP : OA :: DM : DA$$

But $DP = y$, $OA = 2\,a$, $DM = \sqrt{OD \cdot DA} = \sqrt{x\,(2\,a - x)}$,
$DA = 2\,a - x$;

hence $\qquad y : 2\,a :: \sqrt{(2\,a - x)\,x} : 2\,a - x.$

$$\therefore y^2 = \frac{4\,a^2 x}{2\,a - x} \ \cdots \ (1)$$

is the required equation.

Second Method. — Let $MCO = \theta$; then by definition

$$y = 2\,a \tan \frac{\theta}{2} = 2\,a \sqrt{\frac{a\,(1 - \cos \theta)}{a\,(1 + \cos \theta)}}.$$

But $a\,(1 - \cos \theta) = a - a \cos \theta = OC - DC = OD = x$, and
$a\,(1 + \cos \theta) = a + a \cos \theta = OC + DC = OD' = 2\,a - x$;

hence $\qquad y = 2\,a \sqrt{\dfrac{x}{2\,a - x}}\,;$

or, squaring $\qquad y^2 = \dfrac{4\,a^2 x}{2\,a - x}\,.$

This curve was invented by Donna Maria Agnesi, an Italian mathematician of the eighteenth century.

Schol. Solving (1) with respect to y, we have

$$y = \pm\, 2\,a \sqrt{\frac{x}{2\,a - x}}\,.$$

Hence (*a*) the witch is symmetrical with respect to the X-axis.

(*b*) $x = 0$ and $x = 2\,a$ are the equations of its limits.

(*c*) $x = 2\,a$ is the equation of the rectilinear asymptote SS'.

EQUATIONS OF THE FOURTH DEGREE.

153. THE CONCHOID.

The conchoid is the locus generated by the intersection of a circle with a secant line passing through its centre and a fixed point A as the centre of the circle moves along a fixed line OX.

As the intersection of the circle and secant will give two points P, P, one above and the other below the fixed line, it is evident that during the motion of the circle these points will generate a curve with two branches. The upper branch MBM' is called the SUPERIOR BRANCH; the lower, the INFERIOR BRANCH. The radius of the moving circle O'P (= OB) is called the MODULUS. The fixed line OX is called the DIRECTRIX; the point A, the POLE.

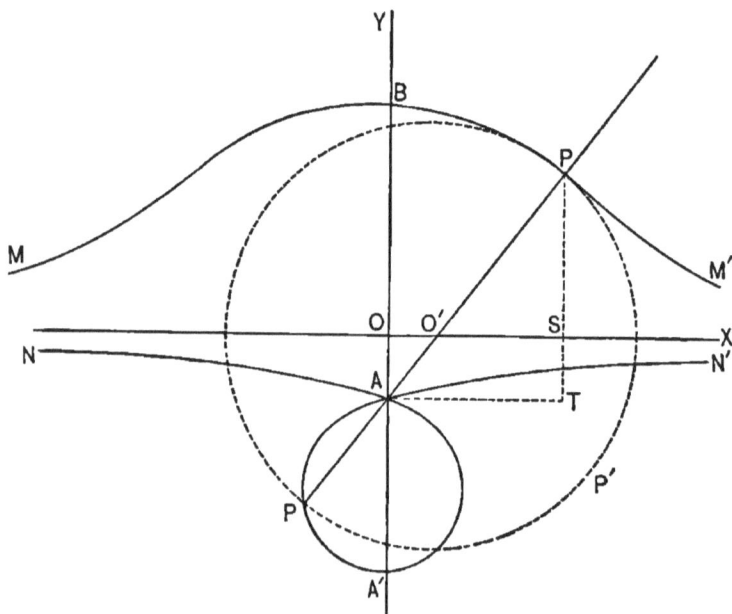

FIG. 63.

1. *To deduce the rectangular equation.*

Let P (x, y), the intersection of the circle PP'P and the

secant AO'P, be any point of the curve. Let $O'P = OB = b$, and let $OA = a$.

The equation of the circle whose centre is at O' $(x', 0)$ is

$$(x - x')^2 + y^2 = b^2.$$

The equation of the line AO'P is

$$y = sx - a \ldots (1)$$

Making $y = 0$ in (1), we have

$$x = \frac{a}{s}$$

for the distance OO'.

But $OO' = x'$; hence

$$\left(x - \frac{a}{s}\right)^2 + y^2 = b^2 \ldots (2)$$

is the equation of the circle. If we now combine (1) and (2) so as to eliminate s, the resulting equation will express the relationship between the co-ordinates of the locus generated by the *intersection* of the loci they represent. Substituting the value of s drawn from (1) in (2), we have

$$\left(x - \frac{ax}{a + y}\right)^2 + y^2 = b^2;$$

$$\therefore x^2 y^2 = (b^2 - y^2)(a + y)^2 \ldots (3)$$

is the required equation.

We might have deduced this equation in the following very simple way: Draw AT ∥ to OX, and PT ∥ to OY. Since the triangles ATP and O'SP are similar, we have

$$PS : SO' :: PT : TA;$$

i.e., $y : \sqrt{b^2 - y^2} :: a + y : x.$

Hence $x^2 y^2 = (b^2 - y^2)(a + y)^2.$

This curve was invented by Nicomedes, a Greek mathematician who flourished in the second century of our era.

It was employed by him in solving the problems of the duplication of a cube and the trisection of an angle.

2. *To deduce the polar equation.*

From the figure we have (AY being the initial line, and A the pole)

$$(AP, PAB) = (r, \theta)$$

But $AP = AO' \pm O'P$;

hence $r = a \sec \theta \pm b$

is the polar equation of the curve.

Schol. Solving (3) with respect to x, we have

$$x = \pm \frac{a + y}{y} \sqrt{b^2 - y^2}.$$

An inspection of this value shows

(*a*) That the conchoid is symmetrical with respect to the Y-axis.

(*b*) That $y = b$ and $y = -b$ are the equations of its limits.

(*c*) That $y = 0$ gives $x = \pm \infty$, ∴ the X-axis is an asymptote.

(*d*) If $a = 0$, then $x = \pm \sqrt{b^2 - y^2}$; i.e., the conchoid becomes a circle.

(*e*) If $b > a$, the inferior branch has a loop as in the figure.

(*f*) If $b = a$, the points A′ and A coincide and the loop disappears.

(*g*) If $b < a$, the inferior branch is similar in form to the superior branch, and the point A $(o, -a)$ is *isolated*; i.e., though entirely separated from the curve, its co-ordinates still satisfy the equation.

154. *To trisect an angle by the aid of the conchoid.*

Let PCX be the angle which we wish to trisect. From C with any radius as CD describe the semi-circle DAH. From the point A draw AB ⊥ to CX and make OB = CD. With A as a pole and OB as a modulus construct a conchoid on CX as a directrix. Join H, the intersection of the inferior branch and the circle, with A and produce it to meet the directrix in K; then

$$CKA = \frac{1}{3} PCX.$$

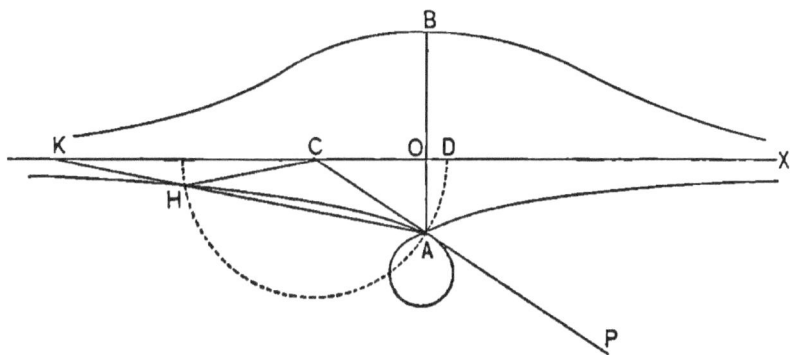

Fɪɢ. 64.

For join H and C; then from the nature of the conchoid

$$HK = HC = OB.$$

From the figure $PCX = CAK + CKA$;

but $CAK = CHA = 2\,CKA$;

hence $PCX = 2\,CKA + CKA.$

Therefore $CKA = \tfrac{1}{3}\,PCX.$

We might have used the superior branch for the same purpose.

155. Tʜᴇ Lɪᴍᴀçᴏɴ.

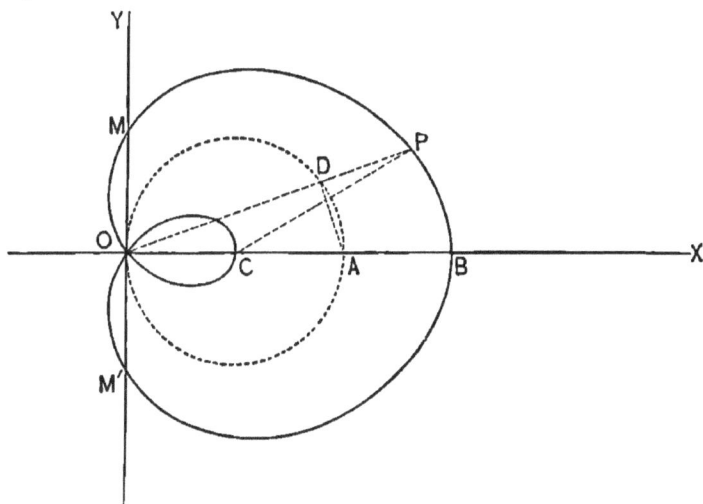

Fɪɢ. 65.

The limaçon is the locus generated by the intersection of two lines OP, CP which are so related that during their revolution about the points O and C the angle PCX is always equal to $\frac{3}{2} POX$.

1. *To deduce the polar equation.*

Let O be the pole, and OX the initial line. Let P be any point of the curve, and let $OC = a$; then

$$(OP,\ POX) = (r,\ \theta).$$

From the triangle POC, we have

$$OP : OC :: \sin OCP : \sin OPC;$$

i.e., $r : a :: \sin \tfrac{3}{2}\theta : \sin \tfrac{1}{2}\theta.$

Hence $r = \dfrac{a \sin \tfrac{3}{2}\theta}{\sin \tfrac{1}{2}\theta}.$

From Trigonometry

$$\sin \tfrac{3}{2}\theta = 3 \sin \tfrac{1}{2}\theta - 4 \sin^3 \tfrac{1}{2}\theta = (3 - 4 \sin^2 \tfrac{1}{2}\theta) \sin \tfrac{1}{2}\theta;$$

hence $r = a\,(3 - 4 \sin^2 \tfrac{1}{2}\theta),$

$$= a\left(3 - 4\,\frac{1 - \cos\theta}{2}\right),$$

$$= a\,(1 + 2\cos\theta) \ \ldots\ (1)$$

is the polar equation of the limaçon.

2. *To deduce the rectangular equation.*

From Art. 35, we have

$$r = \sqrt{x^2 + y^2},\ \cos\theta = \frac{x}{\sqrt{x^2 + y^2}}$$

for the equations of transformation from polar to rectangular co-ordinates. Substituting these values in (1), we have

$$\sqrt{x^2 + y^2} - a = \frac{2\,ax}{\sqrt{x^2 + y^2}}\,;$$

or $(x^2 + y^2 - 2\,ax)^2 = a^2\,(x^2 + y^2) \ \ldots\ (2)$

for the required equation.

Schol. 1. From the triangle ODA, we have

$$OD = OA \cos\theta = 2\,a \cos\theta.$$

From (1) $OP = a + 2\,a \cos\theta;$

hence $OP - OD = DP = a;$

i.e., the intercept between the circle ODA and the limaçon of the secant through O is constant and equal to the radius of the circle.

SCHOL. 2.   If $\theta = 0$, $r = 3\,a = OB$.
If $\theta = 90°$, $r = a = OM$.
If $\theta = 180°$, $r = -a = OC$
If $\theta = 270°$, $r = a = OM'$

156. THE LEMNISCATA.

The lemniscata is the locus generated by the intersection of a tangent line to the equilateral hyperbola with a perpendicular let fall on it from the origin as the point of tangency moves around the curve.

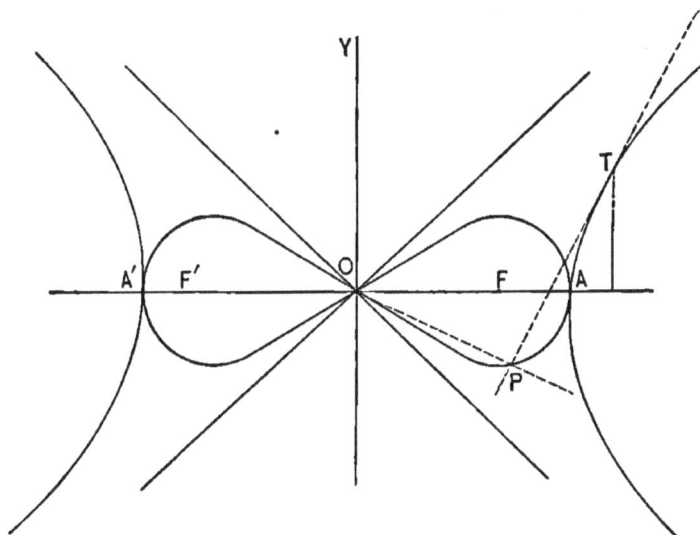

FIG. 66.

1. *To deduce the rectangular equation.*

Since T (x'', y'') is a point of the equilateral hyperbola, we have, Art. 103, Cor 1,

$$x''^2 - y''^2 = a^2 \dots (1)$$

The equation of the tangent line TP is, Art. 112,

$$xx'' - yy'' = a^2 \dots (2)$$

Since the slope of this line is $\dfrac{x''}{y''}$, the equation of the perpendicular OP is

$$y = -\frac{y''}{x''} x \ . \ . \ . \ (3)$$

Treating (2) and (3) as simultaneous and solving for x'' and y'', we find

$$x'' = \frac{a^2 x}{x^2 + y^2} \text{ and } y'' = -\frac{a^2 y}{x^2 + y^2}.$$

Substituting these values in (1), we have

$$\frac{a^4 x^2 - a^4 y^2}{(x^2 + y^2)^2} = a^2 \ ;$$

or . $$(x^2 + y^2)^2 = a^2 (x^2 - y^2) \ . \ . \ . \ (4)$$

for the required equation.

This curve was invented by James Bernouilli. It is *quadrable*, its area being equal to the square constructed on the semi-transverse axis OA.

2. *To deduce the polar equation.*

We have Art. 34, (3), for the equations of transformation

$$x = r \cos \theta, \ y = r \sin \theta.$$

These values in (4) give

$$\{r^2 (\cos^2 \theta + \sin^2 \theta)\}^2 = a^2 \{r^2 (\cos^2 \theta - \sin^2 \theta)\};$$

therefore $$r^4 = a^2 \, r^2 \cos 2 \theta,$$

or . $$r^2 = a^2 \cos 2 \theta \ . \ . \ . \ (5)$$

is the required equation.

SCHOL. If $\theta = 0$, $\cos 2\theta = \cos 0 = 1 \therefore r = \pm a$.

If $\theta < 45°$, $\cos 2 \theta < \cos 90° \therefore r$ has two equal values with opposite signs.

If $\theta = 45°$, $\cos 2 \theta = \cos 90° = 0 \therefore r = 0$.

If $\theta > 45°$ and $<135°$ r is imaginary.

If $\theta = 135°$, $\cos 2 \theta = \cos 270° = 0 \therefore r = 0$.

If $\theta = 180°$, $\cos 2 \theta = \cos 360° = 1 \therefore r = \pm a$.

An examination of these values of r shows that the curve occupies the opposite angles formed by the asymptotes of the hyperbola.

The curve is symmetrical with respect to both axes.

TRANSCENDENTAL EQUATIONS.

157. The Curve of Sines.

This curve takes its name from its equation

$$y = \sin x,$$

and may be defined as a curve whose ordinates are the sines of the corresponding abscissas, the latter being considered as rectified arcs of a circle.

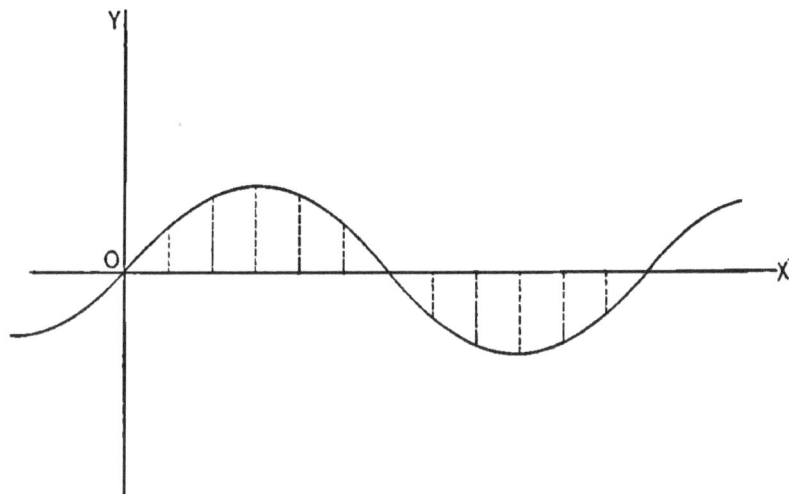

FIG. 67.

To construct the curve. Give values to x which differ from each other by 30°, and find from a "Table of Natural Sines" the values of the corresponding ordinates.

Tabulating the result, we have,

Value of x	Corresponding	Value of y
0	"	0
$30° = \dfrac{\pi}{6} = .52$	"	.50
$60° = \dfrac{2\pi}{6} = 1.04$	"	.87
$90° = \dfrac{3\pi}{6} = 1.56$	"	1.00
$120° = \dfrac{4\pi}{6} = 2.08$	"	.87

Value of x	Corresponding	Value of y
$150° = \dfrac{5\,\pi}{6} = 2.60$	"	.50
$180° = \pi = 3.14$	"	0
$210° = \dfrac{7\,\pi}{6} = 3.66$	"	$-.50$
$240° = \dfrac{8\,\pi}{6} = 4.18$	"	$-.87$
$270° = \dfrac{9\,\pi}{6} = 4.70$	"	-1.00
$300° = \dfrac{10\,\pi}{6} = 5.22$	"	$-.87$
$330° = \dfrac{11\,\pi}{6} = 5.75$	"	$-.50$
$360° = 2\,\pi = 6.28$	"	0

Constructing these points and tracing a smooth curve through them, we have the required locus. As x may have any value from 0 to $\pm\infty$ and yet satisfy the equation of the curve, it follows that the curve itself extends infinitely in the direction of both the positive and negative abscissas.

158. THE CURVE OF TANGENTS.

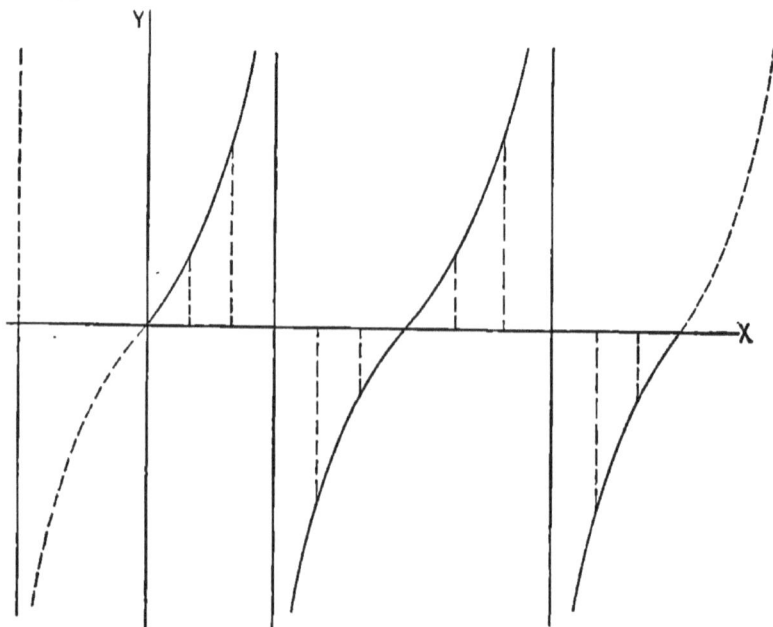

FIG. 68.

This curve also takes its name from its equation
$$y = \tan x.$$

To construct the curve. Give x values differing from each other by 30° and find from a Table of Natural Tangents the corresponding values of y. Tabulating, we have,

Value of x	Corresponding	Value of y
0	"	0
$30° = \frac{\pi}{6} = .52$	"	.57
$60° = \frac{2\pi}{6} = 1.04$	"	1.73
$90° = \frac{3\pi}{6} = 1.56$	"	∞
$120° = \frac{4\pi}{6} = 2.08$	"	-1.73
$150° = \frac{5\pi}{6} = 2.60$	"	$-.57$
$180° = \pi = 3.14$	"	0
$210° = \frac{7\pi}{6} = 3.66$	"	.57
$240° = \frac{8\pi}{6} = 4.18$	"	1.73
$270° = \frac{9\pi}{6} = 4.70$	"	∞
$300° = \frac{10\pi}{6} = 5.22$	"	-1.73
$330° = \frac{11\pi}{6} = 5.75$	"	$-.57$
$360° = 2\pi = 6.28$	"	0

Constructing these points and tracing a smooth curve through them, we have the locus of the equation.

This curve, together with that of the preceding article, belong to the class of *Repeating Curves*, so called because they repeat themselves infinitely along the X-axis.

159. THE CYCLOID.

This curve is the locus generated by a point on the circumference of a circle as the circle rolls along a straight line. The line OM is called the BASE of the cycloid; the

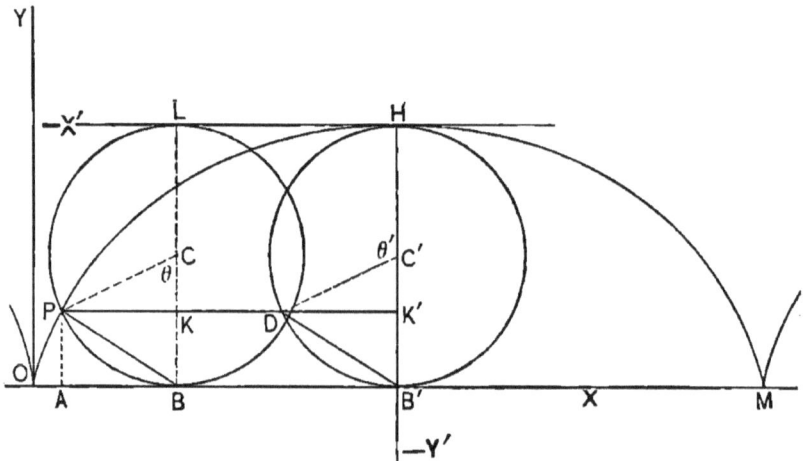

Fig. 69.

point P, the GENERATING POINT; the circle BPL, the GENERATING CIRCLE; the line HB', perpendicular to OM at its middle point, the AXIS. The points O and M are the VERTICES of the cycloid.

1. *To deduce the rectangular equation, the origin being taken at the left-hand vertex of the curve.*

Let P be any point on the curve, and the angle through which the circle has rolled, PCB = θ. Let LB, the diameter of the circle, = 2 a.

Then   OA = OB − AB and AP = CB − CK.

But OA = x, OB = a θ, AB = PK = a sin θ, AP = y, CB = a, CK = a cos θ; hence, substituting, we have

$$x = a\,\theta - a \sin \theta \atop y = a - a \cos \theta \left. \right\} \;\; \cdots \; (1)$$

Eliminating θ between these equations, we have

$$x = a \cos^{-1} \frac{a - y}{a} - \sqrt{2\,ay - y^2} = a \text{ vers}^{-1} \frac{y}{a}$$
$$- \sqrt{2\,ay - y^2} \ldots (2)$$

for the required equation.

Schol. An inspection of (2) shows

(*a*) that negative values of y render x imaginary.

(*b*) When $y = 0$, $x = a \text{ vers}^{-1} 0 = 0$; but $a \text{ vers}^{-1} 0 = 2\,\pi\,a$, or $4\,\pi\,a$, or $6\,\pi\,a$, or etc.; hence there are an infinite number of points such as O and M.

(*c*) When $y = 2\,a$, $x = a \text{ vers}^{-1} 2 = \pi\,a = \text{OB}'$; but $a \text{ vers}^{-1} 2 = 3\,\pi\,a$, or $5\,\pi\,a$, or $7\,\pi\,a$, or etc.; hence, there are an infinite number of points such as H.

(*d*) $y = 0$ and $y = 2\,a$ are equations of the limits.

(*e*) For every value of y between the limits 0 and $2\,a$ there are an infinite number of values for x.

2. *To deduce the rectangular equation, the origin being at the highest point H.*

We have for the equations of transformation

$$x = \text{OA} = \text{OB}' - \text{PK}' = \pi\,a + x'$$
$$y = \text{AP} = \text{B}'\text{H} - \text{HK}' = 2\,a + y'$$

These values in (1) above give

$$\left. \begin{array}{l} x' = a\,(\theta - \pi) - a \sin \theta \\ y' = -\,a - a \cos \theta \end{array} \right\} \ldots (3)$$

But θ', the angle through which the circle has rolled from H, $= \theta - \pi$; hence

$$\left. \begin{array}{l} x' = a\,\theta' + a \sin \theta' \\ y' = a\,(\cos \theta' - 1) \end{array} \right\} \ldots (4)$$

Hence $\qquad x' = a \text{ vers}^{-1} \dfrac{-\,y'}{a} + \sqrt{-\,2\,ay' - y'^2} \ldots (5)$

The invention of this curve is usually attributed to Galileo. With the exception of the conic sections no known curve possesses so many useful and beautiful properties. The following are some of the more important:

1. Area OPHDB'O $=$ area HDB' $= \pi a^2$.
2. Area of cycloid OHMO $= 3$ HDB' $= 3 \pi a^2$.
3. Perimeter OPHM $= 4$ HB' $= 8 a$.
4. If two bodies start from any two points of the curve (the curve being inverted and friction neglected), they will reach the lowest point H at the *same* time.
5. A body rolling down this curve will reach the lowest point H in a *shorter* time than if it were to pursue any other path whatever.

SPIRALS.

160. The SPIRAL is a transcendental curve generated by a point revolving about some fixed point, and receding from it in obedience to some fixed law.

The portion of the locus generated during one revolution of the point is called a SPIRE.

The circle whose radius is equal to the radius-vector of the generating point at the end of the first revolution is called the MEASURING CIRCLE of the spiral.

161. THE SPIRAL OF ARCHIMEDES.

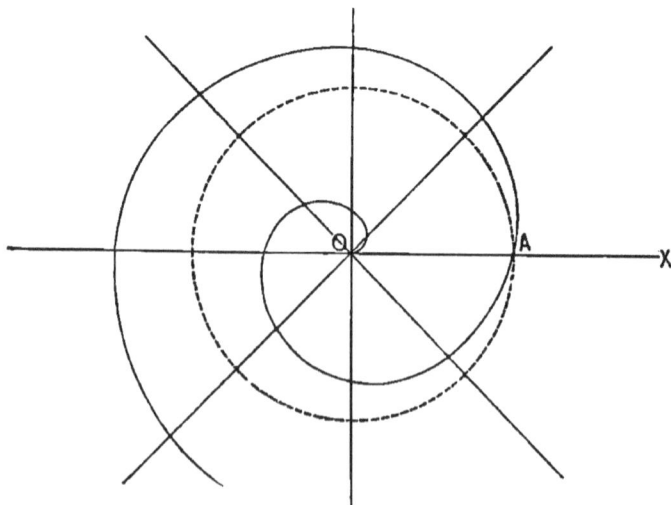

FIG. 70.

This spiral is the locus generated by a point so moving that the ratio of its radius-vector to its vectorial angle is always constant.

From the definition, we have

$$\frac{r}{\theta} = c \; ;$$

hence $\qquad r = c\,\theta \; \ldots \; (1)$

is the equation of the spiral.

To construct the spiral.

Assuming values for θ and finding from (1) the corresponding value for r, we have

Values of θ	Corresponding	Values of r
0	"	0
$45° = \dfrac{\pi}{4}$	"	$\dfrac{\pi}{4}\,c$
$90° = \dfrac{2\,\pi}{4}$	"	$\dfrac{2\,\pi}{4}\,c$
$135° = \dfrac{3\,\pi}{4}$	"	$\dfrac{3\,\pi}{4}\,c$
$180° = \pi$	"	$\pi\,c$
$225° = \dfrac{5\,\pi}{4}$	"	$\dfrac{5\,\pi}{4}\,c$
$270° = \dfrac{6\,\pi}{4}$	"	$\dfrac{6\,\pi}{4}\,c$
$315° = \dfrac{7\,\pi}{4}$	"	$\dfrac{7\,\pi}{4}\,c$
$360° = 2\,\pi$	"	$2\,\pi\,c$
∞	"	∞

Constructing these points and tracing a smooth curve through them, we have a portion of the spiral.

Since $\theta = 0$ gives $r = 0$, the spiral passes through the pole.

Since $\theta = \infty$ gives $r = \infty$, the spiral makes an infinite number of revolutions about the pole.

Since $\theta = 2\,\pi$ gives $r = 2\,\pi\,c$, OA $(= 2\,\pi\,c)$ is the radius of the measuring circle.

162. THE HYPERBOLIC SPIRAL.

This curve is the locus generated by a point so moving that the product of its radius-vector and vectorial angle is always constant.

From the definition we have

$$r\,\theta = c,$$

or
$$r = \frac{c}{\theta} \ \ldots \ (1)$$

for the equation of the spiral.

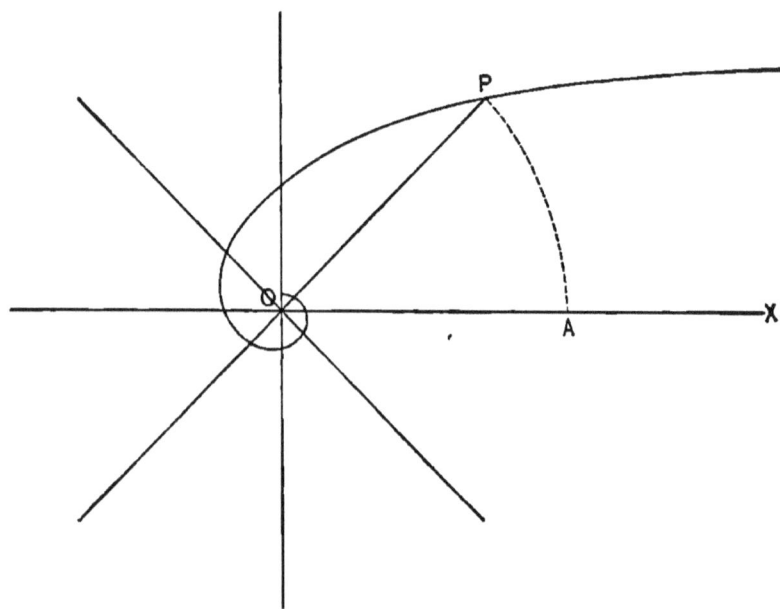

FIG. 71.

To construct the spiral.

Giving values to θ, finding the corresponding values of **r**, we have

Values of θ	Corresponding	Values of r
0	"	∞
$45° = \dfrac{\pi}{4}$	"	$\dfrac{4\,c}{\pi}$

Values of θ	Corresponding	Values of r
$90° = \dfrac{2}{4}\pi$	"	$\dfrac{2c}{\pi}$
$135° = \dfrac{3}{4}\pi$	"	$\dfrac{4c}{3\pi}$
$180° = \pi$	"	$\dfrac{c}{\pi}$
$225° = \dfrac{5}{4}\pi$	"	$\dfrac{4c}{5\pi}$
$270° = \dfrac{6}{4}\pi$	"	$\dfrac{4c}{6\pi}$
$315° = \dfrac{7}{4}\pi$	"	$\dfrac{4c}{7\pi}$
$360° = 2\pi$	"	$\dfrac{c}{2\pi}$
∞	"	0

Constructing the points we readily find the locus to be a curve such as we have represented in the figure.

Since $\theta = 0$ gives $r = \infty$ there is no point of the spiral corresponding to a zero-vectorial angle.

Since $\theta = \infty$ gives $r = 0$, the spiral makes an infinite number of revolutions about the pole before reaching it.

Since $\theta = 2\pi$ gives

$$r = \frac{c}{2\pi},$$

c is the circumference of the measuring circle.

SCHOL. Let P be any point on the spiral; then

$$(\text{OP, POA}) = (r, \theta).$$

With O as a centre and OP as a radius describe the arc PA. By circular measure, Arc $PA = r\theta$, and from (1) $c = r\theta$; hence   Arc $PA = c$;

i.e., the arc of any circle between the initial line and the spiral is equal to the circumference of the measuring circle.

163. THE PARABOLIC SPIRAL.

This spiral is the locus generated by a point so moving that the ratio of the square of its radius-vector to its vectorial angle is always constant.

From the definition we have

$$\frac{r^2}{\theta} = c,$$

or, $\qquad r^2 = c\,\theta \ \ldots \ (1)$

for the equation of the spiral.

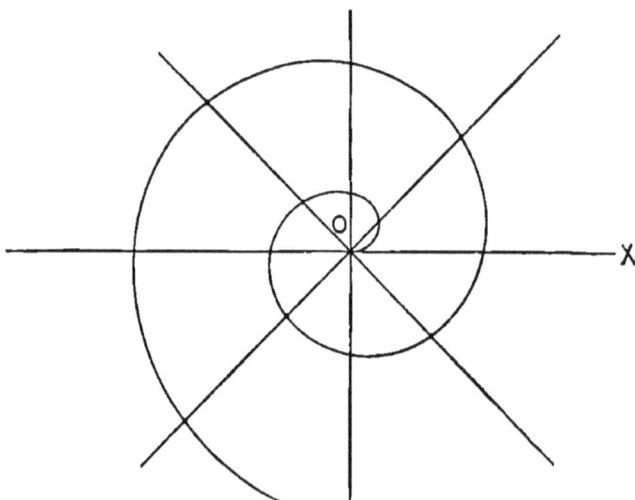

FIG. 72.

To construct the spiral.

Values of θ	Corresponding	Values of r
0	"	0
$45° = \dfrac{\pi}{4}$	"	$\frac{1}{2}\sqrt{c\,\pi}$
$90° = \dfrac{2\,\pi}{4}$	"	$\frac{1}{2}\sqrt{2\,c\,\pi}$
$135° = \dfrac{3\,\pi}{4}$	"	$\frac{1}{2}\sqrt{3\,c\,\pi}$
$180° = \pi$	"	$\sqrt{c\,\pi}$

Values of θ	Corresponding	Values of r
$225° = \dfrac{5\pi}{4}$	"	$\tfrac{1}{2}\sqrt{5\,c\,\pi}$
$270° = \dfrac{6\pi}{4}$	"	$\tfrac{1}{2}\sqrt{6\,c\,\pi}$
$315° = \dfrac{7\pi}{4}$	"	$\tfrac{1}{2}\sqrt{7\,c\,\pi}$
$360° = 2\pi$	"	$\sqrt{2\,c\,\pi}$
∞	"	∞

Constructing these points and tracing a smooth curve through them we have the required locus.

Since $\theta = 0$ gives $r = 0$, the spiral passes through the pole.

Since $\theta = \infty$ gives $r = \infty$, the spiral has an infinite number of spires.

164. THE LITUUS OR TRUMPET.

This curve has for its equation

$$r^2\theta = c,$$

or $$r = \sqrt{\frac{c}{\theta}} \ \cdot \ \cdot \ \cdot \ (1)$$

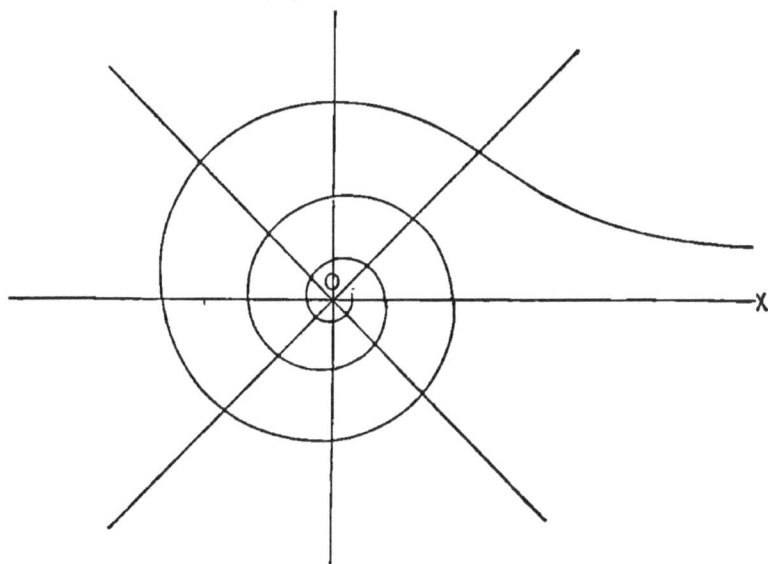

FIG. 73.

If $\theta = 0$, $r = \infty$; if $\theta = \infty$, $r = 0$. This curve has the initial line as an asymptote to its infinite branch.

165. *The Logarithmic Spiral.*

This spiral is the locus generated by a point so moving that the ratio of its vectorial angle to the logarithm of its radius vector is equal to unity.   Hence

$$\frac{\theta}{\log r} = 1 \; ; \text{ i.e., } \theta = \log r;$$

or passing to equivalent numbers (a being the base), we have

$$r = a^{\theta} \; \ldots \; (1)$$

for the equation of the spiral.

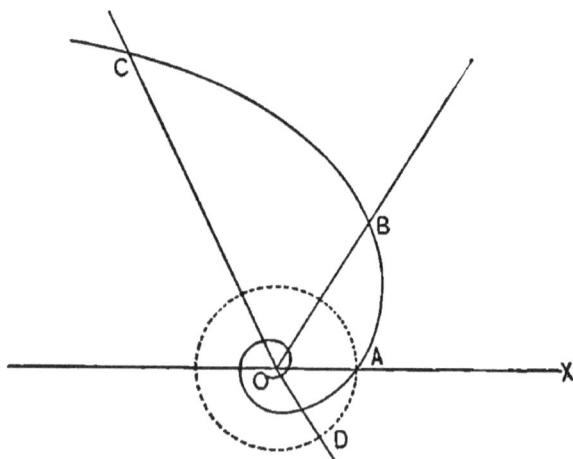

FIG. 74.

To construct the spiral.   Let $a = 2$, then

$$r = 2^{\theta}$$

is the particular spiral we wish to construct.

Values of θ	Corresponding	Values of r
0	"	1
$1 = 57.°3$	"	2
$2 = 114.°6$	"	4
$3 = 171.°9$	"	8
$4 = 229.°2$	"	16
∞	"	∞
$-1 = -57.°3$	"	.5
$-2 = -114.°6$	"	.25
$-3 = -171.°9$	"	.125
$-4 = -229.°2$	"	.062
$-\infty$	"	0

A smooth curve traced through these points will be the required locus.

Since $\theta = 0$ gives $r = 1$ whatever be the assumed value of a, it follows that all logarithmic spirals must intersect the initial line at a unit's distance from the pole.

Since $\theta = \infty$ gives $r = \infty$, the spiral makes an infinite number of revolutions without the circle whose radius OA $= 1$.

Since $\theta = -\infty$ gives $r = 0$, the spiral makes an infinite number of revolutions within the circle OA before reaching its pole.

EXAMPLES.

1. Discuss and construct the cubical parabola

$$y = \frac{x^3}{p^2}.$$

2. What is the polar equation of the limaçon, Fig, 65, the pole being at C ?

$$Ans. \quad r = 2\,a\cos\frac{1}{3}\,\theta.$$

3. Let OF $=$ OF$' = a\sqrt{\tfrac{1}{2}}$, Fig. 66. Show that the lemniscata is the locus generated by a point so moving that the

product of its distances from the two fixed points F, F′ is constant and

$$= \left(\frac{FF'}{2}\right)^2.$$

Discuss and construct the loci of the following equations:

4. $x = \tan y$.

5. $y = \cos x$.

6. $y = \sec x$.

7. $x = \sin y$.

8. $y = \cot x$.

9. $y = \operatorname{cosec} x$.

10. $y = \dfrac{3x - 1}{x^3}$

11. $x^2 y^2 + x y^2 = 1$.

12. $a^3 = x^3 - axy$.

13. $x^{\frac{2}{3}} + y^{\frac{2}{3}} = 1$.

14. $\dfrac{x^2}{a^2} + \dfrac{y^{\frac{2}{3}}}{b^{\frac{2}{3}}} = 1$.

15. $r = a \sin 2\theta$.

16. $r = \dfrac{a}{\sin 2\theta}$.

17. $r = a \sin^3 \dfrac{\theta}{3}$.

18. $r^2 \sin^2 2\theta = 1$.

19. $r = \dfrac{1 + \sin \theta}{1 - \sin \theta}$.

20. Discuss and construct the locus of the equation

$$y^4 - 96\,a^2 y^2 + 100\,a^2 x^2 - x^4 = 0 \text{ or}$$

$$y = \pm \sqrt{48\,a^2 \pm \sqrt{(x - 6\,a)\,(x + 6\,a)\,(x - 8\,a)\,(x + 8\,a)}}.$$

21. Show that $y = \pm x$ are the equations of the rectilinear asymptotes of the locus represented by the equation of Ex. 20.

SOLID ANALYTIC GEOMETRY.

PART II.

CHAPTER I.

CO-ORDINATES. — THE TRI-PLANAR SYSTEM.

166. The position of a point in space is determined when we know its *distance* and *direction* from *three* planes which intersect each other, these distances being measured on lines drawn from the point parallel to the planes. Although it is immaterial in principle what angle these planes make with each other, yet, in practice, considerations of convenience and simplicity have made it usual to take them at right angles. They are so taken in what follows.

Let XOZ, ZOY, YOX be the Co-ordinate Planes intersecting each other at right angles. Let OX, OY, OZ be the Co-ordinate Axes and O, their intersection, the Origin of Co-ordinates.

Let P be any point in the right triedral angle O - XYZ. Then P is completely determined when we know the lengths and directions of the three lines PA, PB, PC let fall from this point on the planes.

As the planes form with each other *eight* right triedral angles, there are evidently *seven* other points which satisfy the condition of being at these distances from the co-ordinate planes. The ambiguity is avoided here (as in the case of the point in a plane) by considering the *directions* in which these lines are measured.

Assuming distances to the *right* of YOZ as *positive*, distances to the *left* will be *negative*.

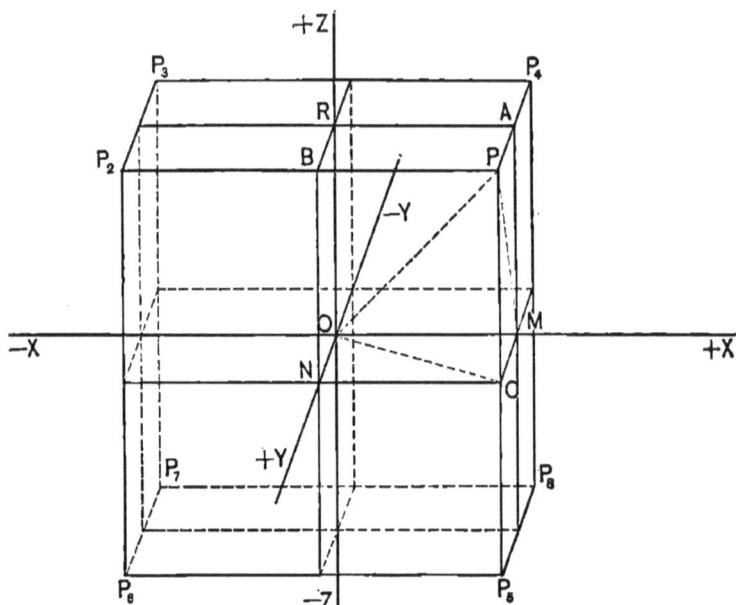

FIG. 75.

Assuming distances *above* XOY as *positive*, distances *below* will be *negative*.

Assuming distances *in front of* XOZ as *positive*, distances *to the rear* will be *negative*.

Calling x', y', z' (= BP, AP, CP, respectively) the *co-ordinates* of the point P in the FIRST ANGLE, we have the following for the co-ordinates of the corresponding points in the other seven:

SECOND ANGLE, above XY plane, to left YZ plane, in front of XZ plane, $(- x', y', z')$ P$_2$.

THIRD ANGLE, above XY plane, to left YZ plane, in rear of XZ plane, $(- x', - y', z')$ P$_3$.

FOURTH ANGLE, above XY plane, to right YZ plane, in rear of XZ plane, $(x', -y', z')$ P$_4$.

FIFTH ANGLE, below XY plane, to right YZ plane, in front of XZ plane, $(x', y', -z')$ P$_5$.

SIXTH ANGLE, below XY plane, to left YZ plane, in front of XZ plane, $(-x', y', -z')$ P$_6$.

SEVENTH ANGLE, below XY plane, to left YZ plane, in rear of XZ plane, $(-x', -y', -z')$ P$_7$.

EIGHTH ANGLE, below XY plane, to right YZ plane, in rear of XZ plane, $(x', -y', -z')$ P$_8$.

EXAMPLES.

1. In what angles are the following points:

$(1, 2, -3)$, $(-1, 3, -2)$, $(-1, -2, -4)$, $(3, -2, 1)$.

2. State the exact position with reference to the co-ordinate axes (or planes) of the following points:

$(0, 0, 2)$, $(-2, 1, 2)$, $(3, 1, 0)$, $(3, -1, 2)$, $(2, 0, 3)$, $(-1, 2, 0)$, $(0, -1, 0)$, $(3, 0, 1)$, $(1, -2, 3)$, $(0, 0, -2)$, $(4, 1, 2)$, $(5, 1, -1)$, $(1, 1, -1)$.

3. In which of the angles are the X-co-ordinates positive? In which negative? In which of the angles are the Y-co-ordinates positive? In which are the Z-co-ordinates negative?

167. Projections. The projection of a point on a plane is the foot of the perpendicular let fall from the point on the plane. Thus A, B, and C, Fig. 75, are the projections of the point P on the planes XZ, YZ, XY, respectively.

The projection of a line of definite length on a plane is the line joining the projections of its extremities on that plane. Thus OC, Fig. 75, is the projection of OP on the XY plane.

The projection of a line of definite length on another line is that portion of the second line included between the feet of the perpendiculars drawn from the extremities of the line of definite length to that line.

Thus OM, Fig. 75, is the projection of OP on the X-axis.

NOTE. — The projections of points and lines as above defined are orthogonal. Unless otherwise stated, all projections will be so understood in what is to follow.

168. *To find the length of a line joining two points in space.*

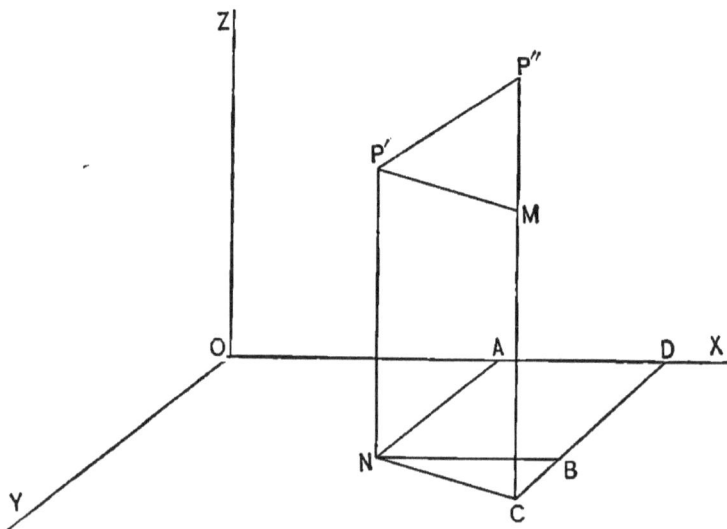

FIG. 76.

Let P' (x', y', z') and P'' (x'', y'', z'') be the given points.

Let L $(=$ P'P''$)$ be the required length. Draw P''C and P'N ‖ to OZ; NA and CD ‖ to OY; NB ‖ to OX. Join N and C and draw P'M ‖ to NC.

We observe from the figure that L is the hypothenuse of a right angled triangle whose sides are P'M and P''M.

Hence

$$L = \sqrt{\overline{P'M}^2 + \overline{P''M}^2} ; \ldots (1)$$

but $\overline{P'M}^2 = \overline{NC}^2 = \overline{NB}^2 + \overline{BC}^2 = (OD - OA)^2 + (DC - AN)^2 =$ $(x'' - x')^2 + (y'' - y')^2$, and $\overline{P''M}^2 = (P''C - P'N)^2 = (z'' - z')^2$.

$$\therefore L = \sqrt{(x'' - x')^2 + (y'' - y')^2 + (z'' - z')^2} \ldots (2)$$

Cor. If $x' = 0$, $y' = 0$, $z' = 0$, then the point P′ coincides with the origin and

$$\therefore L = \sqrt{x''^2 + y''^2 + z''^2} \cdots (3)$$

expresses the distance of a point from the origin.

169. *Given the length and the directional angles of a line joining any point with the origin to find the co-ordinates of the point.*

The Directional angles of a line are the angles which the line makes with the co-ordinate axes.

Let P (x, y, z), Fig. 75, be *any* point, then OP $= L$ will be its distance from the origin. Let POX, POY, POZ $= \alpha, \beta, \gamma$, respectively.

Since OM, ON, OR $(= x, y, z)$ are the projections of OP on X, Y, Z, respectively, we have

$$\left.\begin{array}{l} x = L \cos \alpha \\ y = L \cos \beta \\ z = L \cos \gamma \end{array}\right\} \cdots (1)$$

for the required co-ordinates.

Cor. Squaring and adding equations (1), we have

$$x^2 + y^2 + z^2 = L^2 (\cos^2 \alpha + \cos^2 \beta + \cos^2 \gamma);$$
but $\qquad x^2 + y^2 + z^2 = L^2$ Art. 168 (3);
hence $\qquad \cos^2 \alpha + \cos^2 \beta + \cos^2 \gamma = 1 \cdots (2)$

That is, *the sum of the squares of the directional cosines of a space line is equal to unity.*

Schol. The directional angles of any line, as P′P″, Fig. 76, are the same as those which the line makes with three lines drawn through P′ ∥ to X, Y, Z. The projections of P′P″ on three such lines are $x'' - x'$, $y'' - y'$, $z'' - z'$, Art. 168; hence

$$\left.\begin{array}{l} x'' - x' = L \cos \alpha \\ y'' - y' = L \cos \beta \\ z'' - z' = L \cos \gamma \end{array}\right\} \cdots (3)$$

EXAMPLES.

Required the length of the lines joining the following points :

1. $(1, 2, 3), (- 2, 1, 1),$ **4.** $(0, 0, 0), (2, 0, 1).$
         Ans. $\sqrt{14}.$ *Ans.* $\sqrt{5}.$

2. $(3, - 2, 0), (2, 3, 1).$ **5.** $(0, 4, 1), (- 2, - 1, - 2).$
         Ans. $\sqrt{27}.$ *Ans.* $\sqrt{38}.$

3. $(0, 3, 0), (3, - 1, 0).$ **6.** $(1, - 2, 3), (3, 4, 6).$
         Ans. 5.   *Ans.* 7.

7. Find the distance of the point $(2, 4, 3)$ from the origin ; also the directional cosines of the line.

8. A line makes equal angles with the co-ordinate axes. What are its directional cosines ?

9. Two of the directional cosines of a line are $\sqrt{\tfrac{2}{3}}$ and $\tfrac{1}{3}.$ What is the value of the other ?

10. If (x', y', z') and (x'', y'', z'') are the co-ordinates of the extremities of a line show that

$$\left(\frac{x' + x''}{2}, \ \frac{y' + y''}{2}, \ \frac{z' + z''}{2} \right)$$

are the co-ordinates of its middle point.

THE POLAR SYSTEM.

170. The position of a space point is completely determined when we know its *distance* and *direction* from some *fixed* point. For a complete expression of the *direction* of the point it is necessary that two angles should be given. The angles usually taken are

1st, The angle which the line joining the point and the fixed point makes with a plane passing through the fixed point; and 2d, The angle which the *projection* of the line joining the points on that plane makes with a fixed line in the plane.

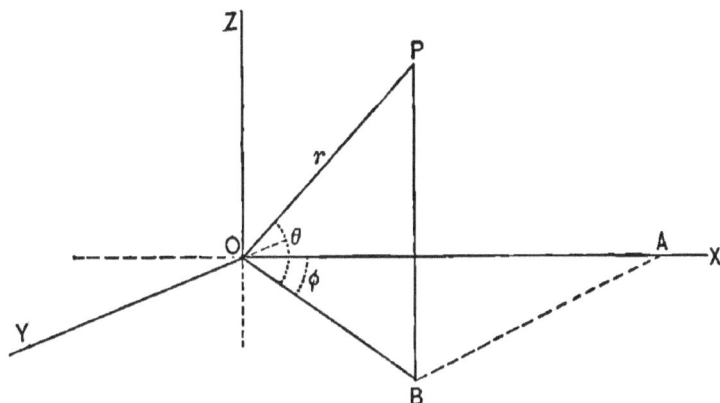

Let O be the fixed point and P the point whose position we wish to determine. Join O and P, and let XOY be any plane passing through O. Let OX be a given line of the plane XOY. Draw PB ⊥ to XOY and pass the plane PBO through PB and OP. The intersection OB of this plane with XOY will be the *projection* of OP on XOY. The angles POB (θ), BOX (φ) and the distance OP (r), when given completely determine the position of P. For the angle φ determines the plane POB, the angle θ determines the line OP in that plane, and the distance r determines the point P on that line.

This method of locating a point is called the POLAR SYSTEM. The angles θ and φ are called VECTORIAL ANGLES, and the distance r is called the RADIUS VECTOR of the point. The point P, when written (r, θ, φ), is said to be expressed in terms of its POLAR CO-ORDINATES.

It is evident by giving all values from 0 to 360° to θ and φ, and all values from 0 to ∞ to r that every point in space may be located.

171. *Given the polar co-ordinates of a point to find its rectangular co-ordinates.*

Draw OY ⊥ to OX and in the plane BOX; draw OZ ⊥ to

OY and OX, and let OX, OY, OZ be the co-ordinate axes. Draw BA ∥ to OY ; then, Fig. 77,
(OA, AB, BP) = (x, y, z) are the rectangular co-ordinates of P.
From the triangle BOP, we have

$$z = r \sin \theta.$$

From the triangle ABO, we have

$$x = OB \cos \varphi.$$

But OB $= r \cos \theta \therefore x = r \cos \theta \cos \varphi.$
From the same triangle we have

$$y = OB \sin \varphi,$$

$\therefore \quad\quad y = r \cos \theta \sin \varphi.$

Hence $\quad\quad \left. \begin{aligned} x &= r \cos \theta \cos \varphi \\ y &= r \cos \theta \sin \varphi \\ z &= r \sin \theta \end{aligned} \right\} \cdots (1)$

express the required relationship.

Cor. If P (x, y, z) be the co-ordinates of *any* point on a locus whose rectangular equation is given then equations (1) are evidently *the equations of transformation from a rectangular system to a polar system, the pole being coincident with the origin.*

Finding the values of r, θ and φ from (1) in terms of x and y, we have

$$\left. \begin{aligned} r &= \sqrt{x^2 + y^2 + z^2} \\ \theta &= \tan^{-1} \frac{z}{\sqrt{x^2 + y^2}} \\ \varphi &= \tan^{-1} \frac{y}{x} \end{aligned} \right\} \cdots (2)$$

for the *equations of transformation from a polar system to a rectangular system, the origin and pole being coincident.*

EXAMPLES.

Find the polar co-ordinates of the following points:

1. $(2, 1, 1)$.   **3.** $(10, 2, 8)$.

2. $(\sqrt{3}, 1, 2\sqrt{3})$.   **4.** $(3, -1, 4)$.

Find the rectangular co-ordinates of the following:

5. $(5, 30°, 60°)$.   **7.** $\left(6, \dfrac{\pi}{3}, \dfrac{\pi}{6}\right)$.

6. $\left(8, \dfrac{\pi}{4}, \dfrac{\pi}{6}\right)$.   **8.** $\left(4, \dfrac{3}{4}\pi, \dfrac{2}{3}\pi\right)$.

Find the polar equations of the following surfaces, the pole and origin being coincident:

9. $x^2 + y^2 + z^2 = a^2$.   *Ans.* $r = a$.

10. $z + sx + ty - c = 0$.

$$Ans. \quad r = \frac{c}{\sin\theta + s\cos\theta\cos\varphi + t\cos\theta\sin\varphi}.$$

Find the directional cosines of the lines joining the following pairs of points:

11. $(1, 2, -1), (3, 2, 1)$.   **13.** $(2, -1, -5), (4, 5, 6)$.

12. $(4, -1, 2), (-1, 3, 2)$.   **14.** $(0, 2, 0), (3, 0, 1)$.

15. If (x', y', z') and (x'', y'', z'') be the co-ordinates of two space points, show that the point

$$\left(\frac{mx'' + nx'}{m+n}, \frac{my'' + ny'}{m+n}, \frac{mz'' + nz'}{m+n} \right)$$

divides the line joining them into two parts which bear to each other the ratio $m : n$.

CHAPTER II.

THE PLANE.

172. *To deduce the equation of the plane.*

Let us assume as the basis of the operation the following property :

If on a perpendicular to a plane two points equidistant from the plane be taken, then every point in the plane is equidistant from these two points, and any point not in the plane is unequally distant.

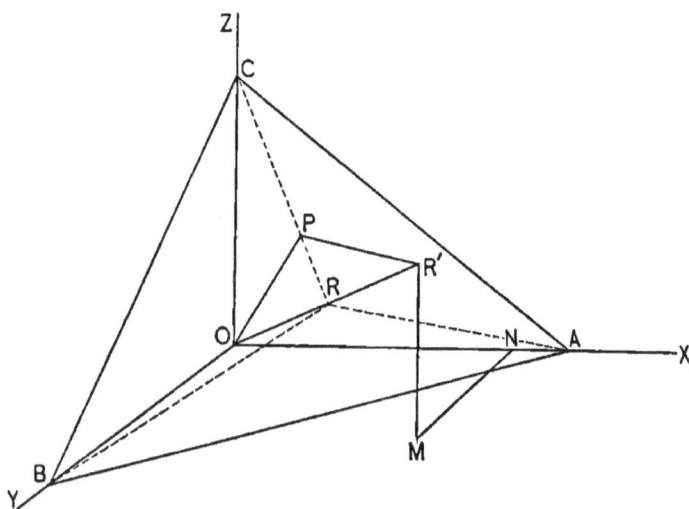

FIG. 78.

Let ABC be any plane. Draw OR ⊥ to ABC, and meeting it in R. Produce OR until $RR' = OR = p$. Every point in the plane is equally distant from O and R'. Let P $(x, y, z,)$ be *any* point of the plane ; let ON, MN, MR', the co-ordinates of $R' = d, e, f$, respectively. Then from Art. 168, (2), we have

$$\overline{PR'}^2 = (d - x)^2 + (e - y)^2 + (f - z)^2$$

From the same article, equation (3), we have

$$OP^2 = x^2 + y^2 + z^2 ;$$

hence, by the assumed property,

$$(d - x)^2 + (e - y)^2 + (f - z)^2 = x^2 + y^2 + z^2.$$

Simplifying this expression, we have

$$dx + ey + fz = \frac{d^2 + e^2 + f^2}{2} \quad \cdots \quad (1)$$

for the required equation.

173. *To find the equation of a plane in terms of the perpendicular to it from the origin and the directional cosines of the perpendicular.*

Let α, β, and γ be the directional angles of the perpendicular $OR' (= 2p)$, Fig. 78. Since ON, MN, MR' $(= d, e, f) =$ the projections of OR' on the co-ordinate axes, we have (Art. 169, (1))

$$\left. \begin{array}{l} d = 2\,p \cos \alpha \\ e = 2\,p \cos \beta \\ f = 2\,p \cos \gamma \end{array} \right\} \quad \cdots \quad (1)$$

Substituting these values in (1), Art. 172, and remembering that $\cos^2 \alpha + \cos^2 \beta + \cos^2 \gamma = 1$, we have

$$x \cos \alpha + y \cos \beta + z \cos \gamma = p \quad \cdots \quad (2)$$

for the required equation. Equation (2) is called the NORMAL EQUATION of the plane.

Since $OR' = 2p = \sqrt{d^2 + e^2 + f^2}$, equations (1) give

$$\cos \alpha = \frac{d}{\sqrt{d^2 + e^2 + f^2}}, \cos \beta = \frac{e}{\sqrt{d^2 + e^2 + f^2}},$$

$$\cos \gamma = \frac{f}{\sqrt{d^2 + e^2 + f^2}}.$$

Substituting these values in (2), we have

$$\frac{d}{\sqrt{d^2 + e^2 + f^2}} x + \frac{e}{\sqrt{d^2 + e^2 + f^2}} y +$$

$$\frac{f}{\sqrt{d^2 + e^2 + f^2}} z = p \ \ldots \ (3)$$

for the equation of the plane expressed in terms of the co-ordinates of a point on the perpendicular to it from the origin and the perpendicular.

COR. 1. If $p = 0$ in (2), we have

$$x \cos \alpha + y \cos \beta + z \cos \gamma = 0 \ \ldots \ (4)$$

for the equation of a plane through the origin.

COR. 2. If $\alpha = 90°$, $\cos \alpha = 0$, hence

$$y \cos \beta + z \cos \gamma = p \ \ldots \ (5)$$

is the equation of a plane $\perp$ to the YZ-plane.
If $\beta = 90°$, we obtain similarly

$$x \cos \alpha + z \cos \gamma = p \ \ldots \ (6)$$

for the equation of a plane $\perp$ to the XZ-plane.
If $\gamma = 90°$, then

$$x \cos \alpha + y \cos \beta = p \ \ldots \ (7)$$

is the equation of a plane $\perp$ to the XY-plane.

COR. 3. If $\alpha = 90°$ and $\beta = 90°$, then

$$z = \frac{p}{\cos \gamma} \ \ldots \ (8)$$

is the equation of a plane $\perp$ to YZ and XZ, and hence $\parallel$ to XY.

Similarly, we find

$$y = \frac{p}{\cos \beta} \ \ldots \ (9)$$

$$x = \frac{p}{\cos \alpha} \ \ldots \ (10)$$

for the equations of planes $\parallel$ to XZ and YZ respectively.

Cor. 4. If $p = 0$ in (8), (9), and (10), then

$$\left.\begin{array}{l} z = 0 \\ y = 0 \\ x = 0 \end{array}\right\} \quad \ldots \text{(11)}$$

are the equations of XY, XZ, and YZ, respectively.

174. *To find the equation of a plane in terms of its intercepts.*

Let, Fig. 78, OA $=a$, OB $= b$, OC $= c$. Since OR $(= p)$ is perpendicular to the plane ABC, we have from the right triangles ORA, ORB, and ORC

$$\left.\begin{array}{l} \cos \alpha = \dfrac{p}{a} \\[2mm] \cos \beta = \dfrac{p}{b} \\[2mm] \cos \gamma = \dfrac{p}{c} \end{array}\right\} \quad \ldots \text{(a)}$$

Substituting these values in the normal equation and reducing, we have

$$\frac{x}{a} + \frac{y}{b} + \frac{z}{c} = 1 \ldots \text{(1)}$$

for the required equation. Equation (1) is called the SYMMETRICAL EQUATION of the plane.

175. *Every equation of the first degree between three variables represents a plane.*

The most general equation of the first degree between three variables is of the form

$$Ax + By + Cz = D \ldots \text{(1)}$$

Dividing both members of this equation by $\sqrt{A^2 + B^2 + C^2}$, we have

$$\frac{A}{\sqrt{A^2 + B^2 + C^2}} x + \frac{B}{\sqrt{A^2 + B^2 + C^2}} y + \frac{C}{\sqrt{A^2 + B^2 + C^2}} z$$

$$= \frac{D}{\sqrt{A^2 + B^2 + C^2}} \ldots \text{(2)}$$

Comparing (2) with (3) of Art. 173, we see that the co-efficients of the variables are the directional cosines of some line expressed in terms of the co-ordinates of one of its points, and that the second member measures the distance of a plane from the origin; hence (2) and therefore (1) is the equation of a plane.

176. *To find the equations of the traces and the values of the intercepts of a plane given by its equation.*

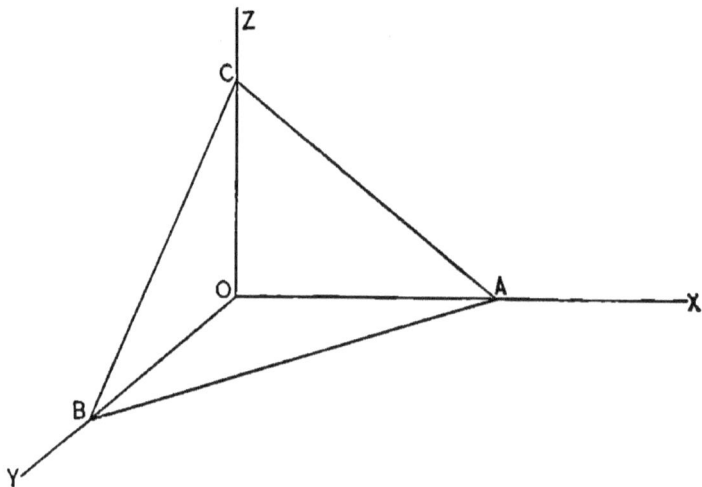

FIG. 70.

Let ABC be the plane and let its equation be

$$Ax + By + Cz = D.$$

1. *To find the equations of the traces AB, BC, AC.*

The traces are the intersections of the given plane with the co-ordinate planes; hence, combining their equations, we have

$$\left. \begin{array}{l} Ax + By + Cz = D \\ z = 0 \end{array} \right\} \therefore Ax + By = D. \quad \text{Trace on XY (AB)} \dots (1)$$

$$\left. \begin{array}{l} Ax + By + Cz = D \\ y = 0 \end{array} \right\} \therefore Ax + Cz = D. \quad \text{Trace on XZ (AC)} \dots (2)$$

$$\left. \begin{array}{l} Ax + By + Cz = D \\ x = 0 \end{array} \right\} \therefore By + Cz = D. \quad \text{Trace on YZ (BC)} \dots (3)$$

2. *To find the intercepts OA, OB, OC.*

The points A, B, C are the intersections of the given plane with the co-ordinate planes taken in pairs; hence, combining their equations, we have

$$\left.\begin{array}{l} Ax + By + Cz = D \\ z = 0 \\ y = 0 \end{array}\right\} \therefore x = \frac{D}{A} = OA \ \ldots \ (4)$$

$$\left.\begin{array}{l} Ax + By + Cz = D \\ x = 0 \\ z = 0 \end{array}\right\} \therefore y = \frac{D}{B} = OB \ \ldots \ (5)$$

$$\left.\begin{array}{l} Ax + By + Cz = D \\ x = 0 \\ y = 0 \end{array}\right\} \therefore z = \frac{D}{C} = OC \ \ldots \ (6)$$

Cor. If the plane is perpendicular to XZ its Y-intercept $= OB = \infty$; hence, equation (5), $B = 0$. Making $B = 0$ in the general equation, we have

$$Ax + Cz = D \ \ldots \ (7)$$

But (7) and (2) are the same equations; hence, *a perpendicular plane and its trace on the plane to which it is perpendicular have the same equation.*

177. *If $x \cos \alpha + y \cos \beta + z \cos \gamma = p$ be the normal equation of a plane, then $x \cos \alpha + y \cos \beta + z \cos \gamma = p \pm d$ is the equation of a parallel plane at the distance d from it.*

For the directional cosines of the perpendiculars are the same; hence, the perpendiculars are coincident; hence, the planes are parallel. The distance of the planes apart is equal to the difference of the perpendiculars drawn to them from the origin; but this difference is $p \pm d - p$; i.e., $\pm d$. Hence, the proposition.

Cor. If (x', y', z') be a point in the plane whose distance from the origin is $p \pm d$; then

$$\pm d = x' \cos \alpha + y' \cos \beta + z' \cos \gamma - p \ \ldots \ (1)$$

is its distance from the parallel plane whose distance from the origin is p. From equations (a), Art. 174, we have

$$\cos \alpha = \frac{p}{a}, \cos \beta = \frac{p}{b}, \cos \gamma = \frac{p}{c};$$

hence $\cos^2 \alpha + \cos^2 \beta + \cos^2 \gamma = \dfrac{p^2}{a^2} + \dfrac{p^2}{b^2} + \dfrac{p^2}{c^2} = 1.$

These values in (1) give

$$\pm d = \left(\frac{x'}{a} + \frac{y'}{b} + \frac{z'}{c} - 1 \right) \frac{abc}{\sqrt{a^2 b^2 + b^2 c^2 + a^2 c^2}} \cdots (2)$$

for the expression of the distance of a point from a plane which is given in its symmetrical form.

Let the student show that the expression for d becomes

$$\pm d = \frac{Ax' + By' + Cz' - D}{\sqrt{A^2 + B^2 + C^2}} \cdots (3)$$

when the equation of the plane is given in its general form.

What is the significance of the double sign in (1), (2), and (3)?

178. *To find the equation of a plane which passes through three given points.*

Let (x', y', z'), (x'', y'', z''), (x''', y''', z''') be the given points. Since the equation we seek is that of a plane, it must be

$$Ax + By + Cz = D \cdots (1)$$

in which A, B, C, D are to be determined by the conditions imposed.

Since the plane is to contain the three given points, the co-ordinates of each of these must satisfy its equation; hence, the following equations of condition:

$$Ax' + By' + Cz' = D$$
$$Ax'' + By'' + Cz'' = D$$
$$Ax''' + By''' + Cz''' = D.$$

These *three* equations contain the *four* unknown quantities A, B, C, D. If we find from the equations the values of A,

B, C in terms of D and the known quantities, and substitute these values in (1), each term of the resulting equation will contain D as a factor. Let

$A = A'D$, $B = B'D$, $C = C'D$ be the values found.

Substituting in (1), we have

$$A'Dx + B'Dy + C'Dz = D.$$
$$\therefore A'x + B'y + C'z = 1 \ldots (2)$$

is the required equation.

179. The preceding discussion has elicited the fact that every equation of the first degree between three variables represents a plane surface. It remains to be shown that every equation between three variables represents a *surface* of some kind.

Let $\qquad z = f(x, y) \ldots (1)$

be any equation between the three variables (x, y, z). Since x and y are independent, we may give them an infinite number of values. For every *pair* of values thus assumed there is a point on the XY plane. These values in (1) give the corresponding value or values of z, which, laid off on the perpendicular erected at the point in the XY plane, will locate one or more points on the locus of the equation. But the number of values of z for any assumed pair of values of x and y are necessarily *finite*, while the number of pairs of values which may be given x and y are *infinite;* hence (1) must represent a surface of some kind.

If

$$\left. \begin{array}{l} z = f(x, y) \\ z = \varphi(x, y) \end{array} \right\} \ldots (2)$$

be the equations of two surfaces, then they will represent their line of intersection if taken *simultaneously.* For these equations can only be satisfied at the same time by the co-ordinates of points common to both. *Hence, in general, two equations between three variables determine the position of a line in space.*

If
$$\left.\begin{array}{l} z = f(x, y) \\ z = \varphi(x, y) \\ z = \psi(x, y) \end{array}\right\} \quad \dots (3)$$

be the equations of three surfaces, then they will represent their *point or points of intersection* when considered as *simultaneous. Hence, in general, three equations between three variables determine the positions of space points.*

EXAMPLES.

Find the traces and intercepts of the following planes:

1. $x - 2y + z = 6.$

2. $\dfrac{3}{4}x - y + \dfrac{z}{2} = 1.$

3. $x - y + 4z = \dfrac{1}{2}.$

4. $2x + 3y - 4z = 0.$

5. $\dfrac{x-1}{2} + \dfrac{y-z}{3} = 2.$

6. $\dfrac{x}{2} - \dfrac{y}{3} + \dfrac{z}{4} = 1.$

7. $\dfrac{x}{3} - \dfrac{y}{2} - \dfrac{z}{4} = 1.$

8. $\dfrac{2x}{3} - \dfrac{3y}{4} + \dfrac{2z}{5} = 1.$

9. $\dfrac{2x}{5} + \dfrac{y}{3} - \dfrac{3z}{4} = 1.$

10. $\dfrac{2x}{y} - 2 = \dfrac{3}{4}.$

11. The directional cosines of a perpendicular let fall from the origin on a plane are $\dfrac{2}{3}, \dfrac{1}{3}, \dfrac{2}{3}$; required the equation of the plane, the length of the perpendicular $= 4.$

$$Ans. \quad \frac{x}{6} + \frac{y}{12} + \frac{z}{6} = 1.$$

Required the equations of the plane whose intercepts are as follows:

12. $1, 2, 3.$

13. $2, -1, 3.$

14. $\dfrac{1}{2}, \dfrac{1}{3}, -2.$

15. $-1, -\dfrac{2}{3}, -4.$

16. What is the equation of the plane, the equations of whose traces are $x - 3y = 4$ and $x + z = 4$?

$$Ans. \quad x - 3y + z = 4.$$

17. The co-ordinates of the projection of a point in the plane $x - 3\,y + 2\,z = 2$ on the XY plane are $(2, 1)$; required the distance of the point from the XY plane.

Ans. $\frac{3}{2}$.

Write the equations of the planes which contain the following points:

18. $(1, 2, 3)$, $(-1, 2, -1)$, $(3, 2, 0)$.

19. $(4, 1, 0)$, $(2, 0, 0)$, $(0, 1, 2)$.

20. $(0, 2, 0)$, $(3, 2, 1)$, $(-1, 0, 2)$.

21. $(2, 2, 2)$, $(3, 3, 3)$, $(-1, -1, -1)$.

Find the point of intersection of the planes

22. $2\,x + y - z = 4$.
$2\,x - 3\,z + y = 10$.
$x + y - z = 2$.

23. $2\,x - y + z = 10$.
$x + y - 2\,z = 3$.
$2\,x - 4\,y + 5\,z = 6$.

24. $2\,x - y - z = 2$.
$2\,x - 3\,y + z = 10$.
$2\,x - y + 2\,z = 8$.

Find the distance of the point $(2, 1, 3)$, from each of the planes

25. $x \cos 60° + y \cos 60° + z \cos 45° = 9$.

26. $x + 3\,y - z = 8$.

27. $x + \dfrac{y}{2} + 3\,z = 4$.

28. $\dfrac{x}{3} - \dfrac{y}{2} + \dfrac{z}{5} = 1$.

29. Find the equation of the plane which contains the point $(3, 2, 2)$ and is parallel to the plane $x - 2\,y + z = 6$.

Reduce the following equations to their normal and symmetrical forms:

30. $2\,x - 3\,y + z = 4$.

31. $4\,x + 2\,y - z = \dfrac{1}{2}$.

32. $\dfrac{2}{3}x + y - \dfrac{1}{4}\,z = 6$.

33. If s, s', s'' represent the sides of the triangle formed by the traces of a plane, and a, b, c represent the intercepts, show that $s^2 + s'^2 + s''^2 = 2\,(a^2 + b^2 + c^2)$.

CHAPTER III.

THE STRAIGHT LINE.

180. *To deduce the equations of the straight line.*

The straight line in space is determined when *two* planes which intersect in that line are given. (See Art. 179.) The equations of any two planes, therefore, may be considered as representing a space line when taken *simultaneously.* Of the infinite number of pairs of planes which intersect in and determine a space line, two of its projecting planes — that is, two planes which pass through the line and are perpendicular to two of the co-ordinate planes — give the simplest equations. For this reason two of these planes are usually selected.

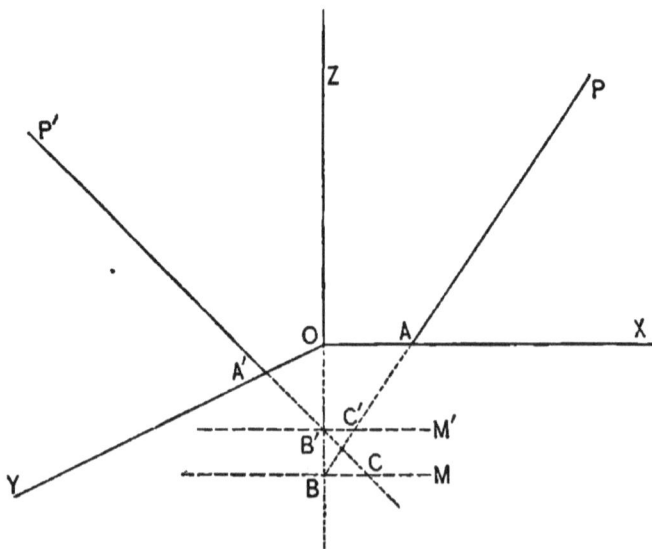

FIG. 80.

Let PBM be the plane which projects a space line on XZ, then its equation will be of the form

$$x = sz + a \text{ (see Art. 176, Cor.)}$$

in which $s = \tan$ ZBP and $a = $ OA.

Let P'B'M' be the plane which projects the line on YZ, then its equation will be

$$y = tz + b,$$

in which $t = \tan$ ZB'P' and $b = $ OA'.

But the two planes determine the line; hence

$$\left. \begin{array}{l} x = sz + a \\ y = tz + b \end{array} \right\} \quad \dots (1)$$

are the required equations.

Cor. 1. If $a = 0$ and $b = 0$, then

$$\left. \begin{array}{l} x = sz \\ y = tz \end{array} \right\} \quad \dots (2)$$

are the equations of a line which pass through the origin.

Cor. 2. If $s = 0$ and $t = 0$, we have

$$\left. \begin{array}{l} x = a \\ y = b \end{array} \right\} \quad \dots (3)$$

for the equation of a line $\parallel$ to the Z-axis.

Cor. 3. Since equations (1) express the relation existing between the co-ordinates of every point on the space line, if we eliminate z from these equations we obtain the immediate relation existing between x and y for points of the line. But this relation is evidently the same for all points in the projecting plane of the line which is $\perp$ to XY and therefore for its trace on XY. But the trace is the projection of the line on XY; hence, eliminating, we have

$$sy - tx = bs - at \dots (4)$$

for the equation of the projection of the line on XY.

181. We have found, Art. 169, Schol., for the length of a line joining two points the expression

$$L = \frac{x'' - x'}{\cos \alpha} = \frac{y'' - y'}{\cos \beta} = \frac{z'' - z'}{\cos \gamma}.$$

Eliminating L and letting x'', y'', z'' ($= x, y, z$) be the co-ordinates of *any* point on the line, we have

$$\frac{x - x'}{\cos \alpha} = \frac{y - y'}{\cos \beta} = \frac{z - z'}{\cos \gamma} \quad \cdots \quad (1)$$

for the SYMMETRICAL EQUATION of a straight line.

182. *To find where a line given by the equations of its projections pierces the co-ordinate planes.*

Let $\left. \begin{array}{l} x = sz + a \\ y = tz + b \end{array} \right\}$ be the equations of the line.

1. *To find where the line pierces the XY-plane.*
The equation of the XY-plane is

$$z = 0.$$

Since the point of intersection is common to both the line and the plane, its co-ordinates must satisfy their equations. Hence

$$x = sz + a$$
$$y = tz + b$$
$$z = 0$$

are simultaneous equations. So treating them we find

$$(a, b, 0)$$

to be the required point.

2. *To find where the line pierces the XZ-plane.*
The equation of the XZ-plane is

$$y = 0.$$

Combining this with the equations of the line, we have

$$\left(\frac{at - sb}{t}, 0, -\frac{b}{t} \right)$$

for the required point.

3. *To find where the line pierces the YZ-plane.*

$$\left. \begin{array}{l} x = sz + a \\ y = tz + b \\ x = 0 \end{array} \right\} \text{ are simultaneous;}$$

hence $\left(0, \frac{sb - at}{s}, -\frac{a}{s} \right)$

is the required point.

183. *To find the equations of a line passing through a given point.*

Let (x', y', z') be the given point.

Since the line is straight its equations are

$$\left. \begin{array}{l} x = sz + a \\ y = tz + b \end{array} \right\} \quad \dots (1)$$

in which the constants are unknown.

Since it is to pass through the point (x', y', z') its equations must be satisfied for the co-ordinates of this point; hence the equations of condition:

$$\left. \begin{array}{l} x' = sz' + a \\ y' = tz' + b \end{array} \right\} \quad \dots (2)$$

As the three conditions imposed by these four equations cannot, in general, be fulfilled by a straight line, we must eliminate one of them. Subtracting the first equation in group (2) from the first in group (1) and the second in group (2) from the second in group (1), we have

$$\left. \begin{array}{l} x - x' = s(z - z') \\ y - y' = t(z - z') \end{array} \right\} \quad \dots (3)$$

for *the general equations of a straight line passing through a point.*

184. *To find the equations of a line passing through two given points.*

Let (x', y', z'), (x'', y'', z'') be the given points.

As the line is straight its equations are

$$\left. \begin{array}{l} x = sz + a \\ y = tz + b \end{array} \right\} \quad \dots (1)$$

in which the constants are to be determined.

As it is to pass through (x', y', z'), we must have

$$\left. \begin{array}{l} x' = sz' + a \\ y' = tz' + b \end{array} \right\} \quad \dots (2)$$

As it is to pass through (x'', y'', z''), we must have also

$$\left. \begin{aligned} x'' &= sz'' + a \\ y'' &= tz'' + b \end{aligned} \right\} \quad \cdots \quad (3)$$

As these six equations impose *four* conditions on the line, we must eliminate two of them. The conditions of the proposition, however, require the line to pass through the two points; hence we must eliminate the other two.

Eliminating a and b from groups (1) and (2), by subtraction, we have

$$\left. \begin{aligned} x - x' &= s\,(z - z') \\ y - y' &= t\,(z - z') \end{aligned} \right\} \quad \cdots \quad (4)$$

Now, eliminating a and b from (2) and (3), we have

$$\left. \begin{aligned} x' - x'' &= s\,(z' - z'') \\ y' - y'' &= t\,(z' - z'') \end{aligned} \right\} \quad \cdots \quad (5)$$

Eliminating s and t between (4) and (5), we have

$$\left. \begin{aligned} x - x' &= \frac{x' - x''}{z' - z''}\,(z - z') \\[2mm] y - y' &= \frac{y' - y''}{z' - z''}\,(z - z') \end{aligned} \right\} \quad \cdots \quad (6)$$

for the required equations.

EXAMPLES.

1. Given the line $\left. \begin{aligned} x &= 2z + 1 \\ y &= 4z - 3 \end{aligned} \right\}$ required the equation of the projection on XY.

$$\textit{Ans.} \quad 2x - y = 5.$$

2. How are the following lines situated with reference to the axes?

$$\left. \begin{aligned} x &= 2 \\ y &= 3 \end{aligned} \right\}, \quad \left. \begin{aligned} y &= 0 \\ z &= 1 \end{aligned} \right\}, \quad \left. \begin{aligned} y &= 0 \\ x &= 1 \end{aligned} \right\}, \quad \left. \begin{aligned} x &= 0 \\ z &= 0 \end{aligned} \right\}, \quad \left. \begin{aligned} x &= 3 \\ y &= 0 \end{aligned} \right\}, \quad \left. \begin{aligned} x &= z \\ y &= 0 \end{aligned} \right\}.$$

Find the co-ordinates of the points in which the following lines pierce the co-ordinate planes:

3. $\left. \begin{aligned} x &= 3z - 1 \\ y &= 2z + 2 \end{aligned} \right\}.$ **4.** $\left. \begin{aligned} x &= -z - 1 \\ y &= 2z + 3 \end{aligned} \right\}.$ **5.** $\left. \begin{aligned} 2x + y &= 3 \\ x - z &= 1 \end{aligned} \right\}.$

6. Given $(2, 1, -2)$, $(3, 0, 2)$; required

(*a*) The length of the line joining the points.

(*b*) The equation of the line.

(*c*) The points in which the line pierces the co-ordinate planes.

Find the equations of the lines which pass through the points:

7. $(2, 1, 3)$, $(3, -1, -1)$.   **9.** $(2, -1, 0)$, $(3, 0, 0)$.

8. $(-1, 2, 3)$, $(-1, 0, 2)$.   **10.** $(1, -1, -2)$, $(-1, -2, -3)$.

11. The projections of a line on XZ and YZ make angles of 45° and 30° respectively with the Z-axis, and the line in space contains the point $(1, 2, 3)$; required the equations of the line.

$$\text{Ans.} \quad \left. \begin{array}{l} x = z - 2. \\ y = \dfrac{z}{\sqrt{3}} - \sqrt{3} + 2. \end{array} \right\}$$

12. The vertices of a triangle are $(2, 1, 3)$, $(3, 0, -1)$, $(-2, 4, 3)$; required the equations of its sides.

13. Is the point $(2, -1, 3)$ on the line which passes through $(-1, 3, 2)$, $(3, 2, -2)$?

14. Write the equations of a line which lies in the plane $x - 2y + 3z = 1$.

NOTE. — Assume two points in the plane; the line joining them will be a line of the plane.

15. Find the equation of a line through $(1, -2, 2)$ which is parallel to the plane $x - y + z = 4$.

16. Find the point in which the line $\left. \begin{array}{l} x + 2z = 3 \\ y - z + 2 = 0 \end{array} \right\}$ pierces the plane $3x + 2y - z = 4$.

17. Required the equation of the plane which contains the two lines $\left. \begin{array}{l} x - 2z - 1 = 0 \\ y - 2z - 2 = 0 \end{array} \right\}$ and $\left. \begin{array}{l} x - z - 5 = 0 \\ y - 4z + 6 = 0 \end{array} \right\}$

18. Find the point of intersection of the planes

$$x + 3y - z = 4,\ x - y + z = 2,\ 2x + y = 3.$$

19. Find the equations of the projecting planes of the line

$$\left.\begin{array}{l} x - 2y + z = 4 \\ 2x + 3y - z = 6 \end{array}\right\}.$$

20. Which angles do the following planes cross ?

$$x - y + z = 4,\ 2x + y - 3z = 2,\ x - 2y - z = 1.$$

185. *To find the intersection of two lines given by their equations.*

Let
$$\left.\begin{array}{l} x = sz + a \\ y = tz + b \end{array}\right\} \text{ and } \left.\begin{array}{l} x = s'z + a' \\ y = t'z + b' \end{array}\right\}$$

be the given equations. Since the point of intersection is common to both lines, its co-ordinates must satisfy their equations. Hence these equations are simultaneous. But we observe that there are *four* equations and only *three* unknown quantities; hence, in order that these equations may consist (and the lines intersect), a certain relationship must exist between the constants which enter into them. To find this relationship, we eliminate x between the *first* and *third*, y between the *second* and *fourth*, and z between the two equations which result. We thus obtain

$$(s - s')\,(b - b') - (t - t')\,(a - a') = 0$$

for the required equation of condition that the two lines shall intersect. If this condition is satisfied for any pair of assumed lines the lines will intersect, and we obtain the co-ordinates of this point by treating any three of the four equations which represent them as simultaneous. So treating the *first, second,* and *third* we obtain

$$\left(\frac{sa' - s'a}{s - s'}, \ t\frac{a' - a}{s - s'} + b, \ \frac{a' - a}{s - s'} \right)$$

for the co-ordinates of the required point.

Note. — We were prepared to expect that our analysis would lead to some conditional equation, for in assuming the equations of two space lines it would be an exceptional case if we so assumed them that the lines which they represent intersected. Lines may cross each other under any angle in space without intersecting. In a plane, however, all lines except parallel lines must intersect. Hence, no conditional equation arose in their discussion.

186. *To find the angle between two lines, given by their equations, in terms of functions of the angles which the lines make with the axes.*

$$\text{Let} \quad \begin{aligned} x &= sz + a \\ y &= tz + b \end{aligned} \right\} \text{ and } \begin{aligned} x &= s'z + a' \\ y &= t'z + b' \end{aligned} \right\}$$

be the equations of the two lines. The angle under which two space lines cross each other is measured by the angle formed by two lines drawn through some point parallel to their directions.

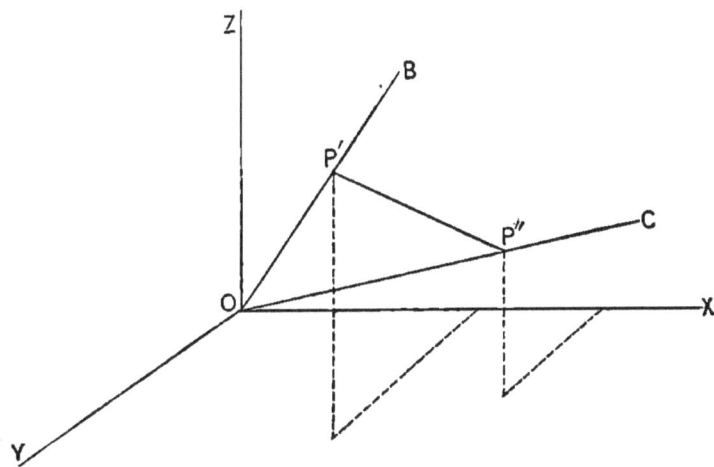

Fig. 81.

Let OB and OC be two lines drawn through the origin parallel to the given lines. Then

$$\begin{aligned} x &= sz \\ y &= tz \end{aligned} \right\} \text{ and } \begin{aligned} x &= s'z \\ y &= t'z \end{aligned} \right\}$$

will be their equations. The angle between these lines is the angle sought. Let φ (= BOC) be this angle.

Let α', β', γ' represent the angles which the line BO makes with X, Y, Z, respectively; and α'', β'', γ'' the angle which CO makes with the same axes. Take any point P' (x', y', z') on OB and any point P'' (x'', y'', z'') on CO and join them by a right line forming the triangle P'OP''.

Let OP' = L', OP'' = L'', and P'P'' = L.

From the triangle P'OP'', we have

$$\cos \varphi = \frac{L'^2 + L''^2 - L^2}{2\,L'L''} \cdots (1)$$

But Art. 168, equation (3) and (2)

$$L'^2 = x'^2 + y'^2 + z'^2,$$
$$L''^2 = x''^2 + y''^2 + z''^2,$$
$$L^2 = (x'' - x')^2 + (y'' - y')^2 + (z'' - z')^2$$
$$= x''^2 + y''^2 + z''^2 + x'^2 + y'^2 + z'^2 - 2\,(x'x'' + y'y'' + z'z'').$$

Substituting these values in (1), we have

$$\cos \varphi = \frac{x'x'' + y'y'' + z'z''}{L'L''} \cdots (2)$$

But Art. 169, (1)

$$x' = L' \cos \alpha', \; y' = L' \cos \beta', \; z' = L' \cos \gamma'$$
$$x'' = L'' \cos \alpha'', \; y'' = L'' \cos \beta'', \; z'' = L'' \cos \gamma''$$

Substituting in (2), we have

$$\cos \varphi = \cos \alpha' \cos \alpha'' + \cos \beta' \cos \beta'' + \cos \gamma' \cos \gamma'' \cdots (3)$$

for the required relation.

Cor. If $\varphi = 90°$

$$\cos \alpha' \cos \alpha'' + \cos \beta' \cos \beta'' + \cos \gamma' \cos \gamma'' = 0 \cdots (4)$$

187. *To find the angle which two space lines make with each other in terms of functions of the angles which the projections of the lines make with the co-ordinate axes.*

Let
$$\begin{array}{l} x = sz \\ y = tz \end{array} \Big\} \text{ and } \begin{array}{l} x = s'z \\ y = t'z \end{array} \Big\}$$

be, as in the preceding article, the equations of the lines

drawn through the origin parallel to the given lines. Since P′ (x', y', z'), Fig. 81, is a point on the first line, we have

$$x' = sz'$$
$$y' = tz',$$

and, Art. 168, $L'^2 = x'^2 + y'^2 + z'^2.$

Eliminating, we find

$$x' = \frac{sL'}{\sqrt{1 + s^2 + t^2}}, \; y' = \frac{tL'}{\sqrt{1 + s^2 + t^2}}, \; z' = \frac{L'}{\sqrt{1 + s^2 + t^2}};$$

and since P″ (x'', y'', z'') is a point on the second line, we have

$$x'' = s'z''$$
$$y'' = t'z'',$$

and, Art. 168, $L''^2 = x''^2 + y''^2 + z''^2.$

Hence,

$$x'' = \frac{s'L''}{\sqrt{1 + s'^2 + t'^2}}, \; y'' = \frac{t'L''}{\sqrt{1 + s'^2 + t'^2}}, \; z'' = \frac{L''}{\sqrt{1 + s'^2 + t'^2}}.$$

But, Art. 169,

$$\cos \alpha' = \frac{x'}{L'} = \frac{s}{\sqrt{1 + s^2 + t^2}}, \; \cos \alpha'' = \frac{x''}{L''} = \frac{s'}{\sqrt{1 + s'^2 + t'^2}}.$$

$$\cos \beta' = \frac{y'}{L'} = \frac{t}{\sqrt{1 + s^2 + t^2}}, \; \cos \beta'' = \frac{y''}{L''} = \frac{t'}{\sqrt{1 + s'^2 + t'^2}}.$$

$$\cos \gamma' = \frac{z'}{L'} = \frac{1}{\sqrt{1 + s^2 + t^2}}, \; \cos \gamma'' = \frac{z''}{L''} = \frac{1}{\sqrt{1 + s'^2 + t'^2}}.$$

Substituting these values in equation (3), Art. 186, and reducing, we have

$$\cos \varphi = \pm \frac{1 + ss' + tt'}{\sqrt{1 + s^2 + t^2} \; \sqrt{1 + s'^2 + t'^2}} \; \cdots (1)$$

for the required expression.

Cor. 1. If $\varphi = 0$, the lines are parallel and equation (1) becomes

$$1 = \pm \frac{1 + ss' + tt'}{\sqrt{1 + s^2 + t^2} \; \sqrt{1 + s'^2 + t'^2}}.$$

Clearing of fractions and squaring, we have

$$(1 + s^2 + t^2)(1 + s'^2 + t'^2) = (1 + ss' + tt')^2.$$

Performing the operations indicated, transposing and collecting, we have

$$(s' - s)^2 + (t' - t)^2 + (st' - s't)^2 = 0.$$

But the sum of the squares of these quantities cannot be equal to zero unless each separately is equal to zero; hence

$$s' = s, \; t' = t, \; st' = s't \ldots (2)$$

are the *conditions for parallelism of space lines.* The first two of these conditions show that if two lines in space are parallel, then their projections on the co-ordinate planes are parallel also. The third condition $(st' = s't)$ is a mere consequence of the other two, and may be omitted in stating the conditions for parallelism.

Cor. 2. If $\varphi = 90°$, the lines are perpendicular to each other, and equation (1) becomes

$$0 = \frac{1 + ss' + tt'}{\sqrt{1 + s^2 + t^2} \; \sqrt{1 + s'^2 + t'^2}} \; ;$$

hence $\qquad 1 + ss' + tt' = 0 \ldots (3)$

is the *condition for perpendicularity in space.*

188. Since the angle which a line makes with any one of the co-ordinate axes is the complement of the angle which the line makes with the co-ordinate plane to which that axis is perpendicular if we let α, β, γ be the complements of α', β', γ', respectively, we have

$$\sin \alpha = \frac{s}{\sqrt{1 + s^2 + t^2}}, \; \sin \beta = \frac{t}{\sqrt{1 + s^2 + t^2}},$$

$$\sin \gamma = \frac{1}{\sqrt{1 + s^2 + t^2}} \ldots (1)$$

for the sines of the angles which a space line makes with the co-ordinate planes.

TRANSFORMATION OF CO-ORDINATES.

189. *To find the equations of transformation from one system of co-ordinates to a parallel system, the origin being changed.*

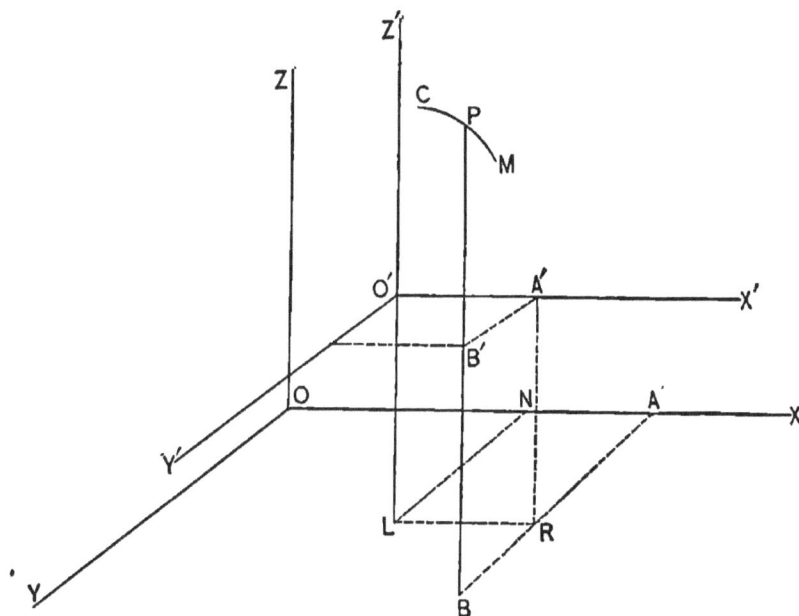

FIG. 82.

Let X, Y, Z be the old axes and X′, Y′, Z′ the new.

Let P be any point on the locus CM. Draw PB, A′R, O′L ∥ to OZ and meeting XOY in B, R and L. Draw BR and produce it to A; BR will be ∥ to OY; draw LN ∥ to BR and LR ∥ to OX. Then (OA, AB, BP) = (x, y, z) are the old co-ordinates of the point P.

(O′A′, A′B′, B′P) = (x′, y′, z′) are the new co-ordinates of the point P.

(ON, NL, LO′) = (a, b, c) are the old co-ordinates of the new origin O′.

From the figure

OA $=$ ON $+$ O'A', AB $=$ NL $+$ A'B', BP $=$ LO' $+$ B'P ;
hence $\qquad x = a + x', y = b + y', z = c + z'$
are the required equations.

190. *To find the equations of transformation from a rectangular system in space to an oblique system, the origin being the same.*

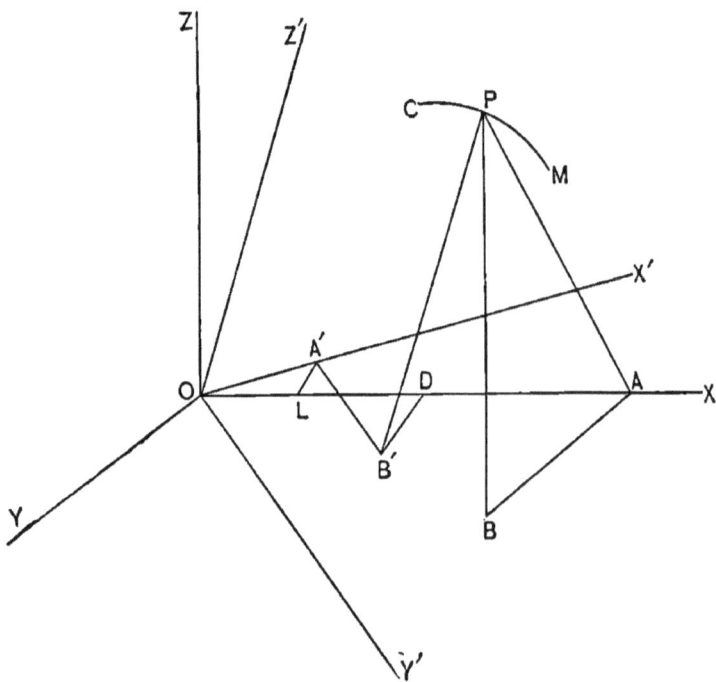

Fig. 83.

Let OX, OY, OZ be the old axes, and OX', OY', OZ' the new.

Let α', β', γ' be the angles which OX' makes with OX, OY, OZ respectively.

Let α'', β'', γ'' be the angles which OY' makes with OX, OY, OZ respectively.

Let α''', β''', γ''' be the angles which OZ' makes with OX, OY, OZ respectively.

Let P be any point on the locus CM. Draw PB and PB'

‖ to OZ and OZ′, respectively, and let B and B′ be the points in which these lines pierce the planes XOY and X′OY′. Draw B′A′ ‖ to OY′; then

(OA, AB, BP) = (x, y, z) are the old co-ordinates of the point P.

(OA′, A′B′, B′P) = (x', y', z') are the new co-ordinates of the point P.

From P, B′ and A′ let fall the perpendiculars PA, B′D, A′L on the X-axis; then from the figure, we have

$$OA = OL + LD + DA$$

But OL, LD and DA are the projections of OA′, A′B′, and PB′, respectively on the X-axis, and each, therefore, is equal to the line whose projection it is into the cosine of the angle which that line makes with the X-axis. (See Art. 169 (1) .)

∴ OL = OA′ cos α', LD = A′B′ cos α'', DA = B′P cos α'''
i.e., OL = x' cos α', LD = y' cos α'', DA = z' cos α''';

hence, substituting, we have

$$\left.\begin{array}{l} x = x' \cos \alpha' + y' \cos \alpha'' + z' \cos \alpha''' \\ \text{Similarly} \quad y = x' \cos \beta' + y' \cos \beta'' + z' \cos \beta''' \\ z = x' \cos \gamma' + y' \cos \gamma'' + z' \cos \gamma''' \end{array}\right\} \quad \ldots (1)$$

Of the nine angles involved in these equations, six only are independent, for since the old axes are rectangular, we must have (See Art. 169, equation (2)).

$$\left.\begin{array}{l} \cos^2 \alpha' + \cos^2 \beta' + \cos^2 \gamma' = 1 \\ \cos^2 \alpha'' + \cos^2 \beta'' + \cos^2 \gamma'' = 1 \\ \cos^2 \alpha''' + \cos^2 \beta''' + \cos^2 \gamma''' = 1 \end{array}\right\} \quad \ldots (2)$$

Cor. 1. If we suppose the new axes to be rectangular also we must have in addition to equation (2) the following conditional equations: See Art. 186, Cor.

$$\left.\begin{array}{l} \cos \alpha' \cos \alpha'' + \cos \beta' \cos \beta'' + \cos \gamma' \cos \gamma'' = 0 \\ \cos \alpha' \cos \alpha''' + \cos \beta' \cos \beta''' + \cos \gamma' \cos \gamma''' = 0 \\ \cos \alpha'' \cos \alpha''' + \cos \beta'' \cos \beta''' + \cos \gamma'' \cos \gamma''' = 0 \end{array}\right\} \quad \ldots (3)$$

Hence, in this case, only *three* of the nine angles involved in equation (1) are independent.

THE CONIC SECTIONS.

191. *The* Conic Sections, *or, more simply,* The Conics, *are the curves cut from the surface of a right circular cone by a plane.*

We wish to show that every such section is an ellipse, a parabola, an hyperbola, or one of their limiting cases. Art. 146.

192. *To deduce the equation of the conic surface.*

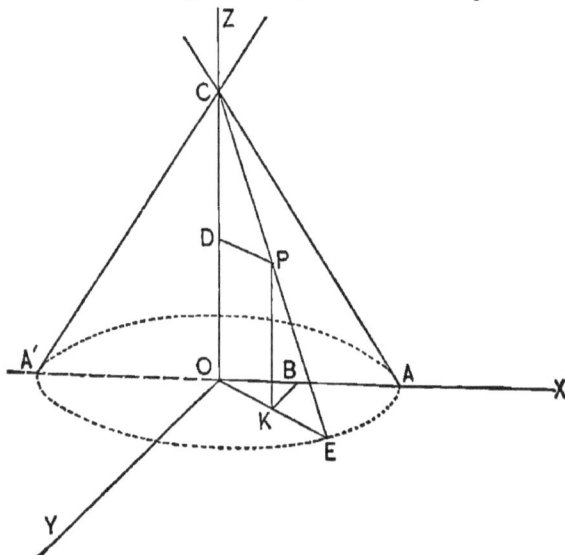

FIG. 84.

Let CAEA′C be the conic surface, generated by revolving the element CA about OZ as an axis. Let P be *any* point on *any* element as CE; let OC = *c* and OEC = *θ*.

Draw DP ∥ to XY-plane and intersecting OZ in D; draw PK ∥ to OZ, KB ∥ to OY, and join O and K producing it to meet the base circle in E.

Then (OB, BK, KP) = (*x, y, z*) are the co-ordinates of P.

From the similar triangles COE, CDP, we have

$$\frac{DC}{DP} = \frac{OC}{OE} = \tan\ \theta \ . \ . \ . \ (1)$$

But $DC = OC - PK = c - z$, and $DP = OK = \sqrt{x^2 + y^2}$; hence,

$$\frac{c - z}{\sqrt{x^2 + y^2}} = \tan \theta;$$

i.e., $\qquad (c - z)^2 = (x^2 + y^2) \tan^2 \theta \;\ldots\; (2)$

is the required equation.

193. *To find the equation of the intersection of a right circular cone and a plane.*

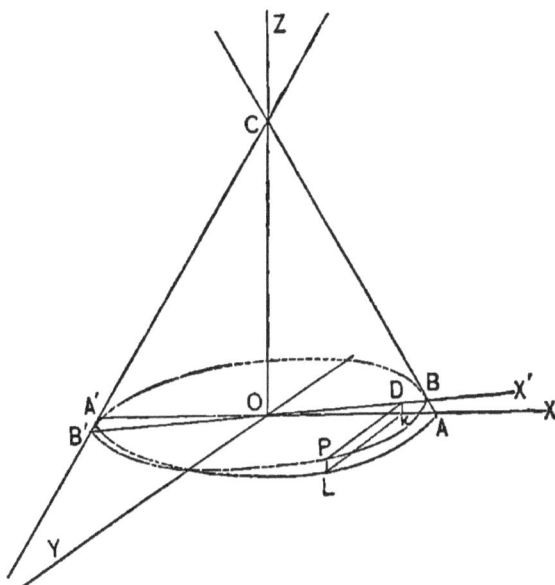

Fig. 85.

Let $CALA'$ be the cone and $X'OY$ the cutting plane. Let $X'OX$, the angle which the cutting plane makes with the plane of the cone's base, $= \varphi$.

Let P (x, y, z) be *any* point on the curve of intersection BPB'. We wish to find the equation of this curve when referred to OY, OX' as axes.

Draw PD ∥ to OY ; PL and DK ∥ to OZ; then

(OK, KL, LP) $= (x, y, z)$ are the space co-ordinates of P,

and (OD, DP) = (x', y) are the co-ordinates of P when referred to OX', OY.

From the figure KL = DP, PL = KD = OD sin φ, OK = OD cos φ.

i.e., $y = y, z = x' \sin \varphi, x = x' \cos \varphi.$

But these values of x, y, z must subsist together with the equation of the conic surface for every point on the curve of intersection; hence substituting in (2), Art. 192, reducing and remembering that $\sin^2 \varphi = \cos^2 \varphi \tan^2 \varphi$, we have, dropping accents,

$$y^2 \tan^2 \theta + x^2 \cos^2 \varphi (\tan^2 \theta - \tan^2 \varphi) + 2cx \sin \varphi - c^2 = 0 \ldots (1)$$

for the equation of the intersection.

By giving every value to φ from 0 to 90° and to c every value from 0 to ∞, equation (1) can be made to represent every section cut from a cone by a plane *except* sections made by planes that are parallel to the co-ordinate planes.

Cor. 1. Comparing (1) with (1), Art. 138, we find

$$\left.\begin{array}{l} a = \tan^2 \theta \\ b = 0 \\ c = \cos^2 \varphi (\tan^2 \theta - \tan^2 \varphi) \end{array}\right\} \ldots (2)$$

Hence, equation (1) represents an ellipse, a parabola, an hyperbola or one of their limiting cases according as, Art. 146.

$$b^2 < 4\,ac$$
$$b^2 = 4\,ac$$
$$b^2 > 4\,ac.$$

Case 1. $\theta > \varphi$. We find this supposition in (2) gives $a > 0$ and $c > 0$; hence, $b^2 < 4\,ac$, i.e., the intersection is an ellipse.

If $\theta > \varphi$ and $c = 0$, the equation resulting from introducing this supposition in (1) can only be satisfied by the point $(0, 0)$; hence it is the equation of two imaginary lines intersecting at the origin.

If $\varphi = 0$, equation (1) becomes

$$y^2 \tan^2 \theta + x^2 \tan^2 \theta = c^2,$$

that is, the intersection is a circle.

CASE 2. $\theta = \varphi$. This supposition in (2) gives $a > 0$ and $c = 0 \therefore b^2 = 4\,ac$. Hence the intersection is a parabola.

If $\theta = \varphi$ and $c = 0$. From (1), we have

$$y^2 \tan^2 \theta = 0; \text{ i.e., } y = 0$$

which is the equation of the X-axis — a straight line.

If $\theta = \varphi = 90°$ and $c = \infty$, then the cone becomes a cylinder, and the cutting plane is perpendicular to its base. The intersection is therefore two parallel lines.

CASE 3. $\theta < \varphi$. This supposition makes $a > 0,\ c < 0$ $\therefore b^2 > 4\,ac$. Hence the intersection is an hyperbola.

If $\theta < \varphi$ and $c = 0$ then (1) becomes

$$y^2 \tan^2 \theta = x^2 \cos^2 \varphi (\tan^2 \varphi - \tan^2 \theta)$$

which is the equation of two intersecting lines.

CASE 4. *Planes* ‖ *to the co-ordinate planes.*

(*a*) *Plane* ‖ *to XY-plane.* Let $z = m$ be the equation of such a plane. Combining it with the equation of the conic surface, we have

$$x^2 + y^2 = \frac{(c - m)^2}{\tan^2 \theta} \ \cdots \ (3)$$

which is the equation of a circle for all values of m.

(*b*) *Plane* ‖ *to YZ-plane.* Let $x = n$ be the equation of such a plane. Combining with (2), Art. 192, we have

$$(c - z)^2 = (n^2 + y^2) \tan^2 \theta$$

or $\qquad y^2 \tan^2 \theta - z^2 + 2\,cz + n^2 \tan^2 \theta - c^2 = 0 \ \cdots \ (4)$

which, since $b^2 > 4\,ac$, is the equation of an hyperbola for all values of n.

(*c*) *Plane* ‖ *to XZ-plane.* Let $y = p$ be the equation of such a plane. Combining with (2), Art. 192, we have after reduction

$$x^2 \tan^2 \theta - z^2 + 2\,cz + p^2 \tan^2 \theta - c^2 = 0 \ \cdots \ (5)$$

which, since $b^2 > 4\,ac$, is the equation of an hyperbola for all values of p.

Hence, in all possible positions of the cutting plane, *the intersection is an ellipse, a parabola, an hyperbola, or one of their limiting cases.*

Note. — Equations (3), (4), (5) of case 4 are the equations of the *projections* of the curves of intersection on the planes to which they are parallel. But the projection of any plane curve on a parallel plane is a curve equal to the given curve; hence the conclusions of case 4 are true for the curves themselves.

194. We have defined the conics, Art. 191, as the curves cut from the surface of a right circular cone by a plane, and assuming this definition we have found and discussed their general equation, Art. 193.

A conic, however, may be otherwise defined as the locus generated by a point so moving in a plane that the ratio of its distance from a fixed point and a fixed line is always constant.

195. *To deduce the general equation of a conic.*

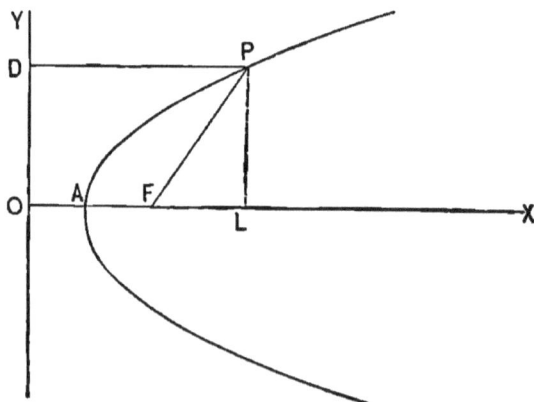

Fig. 86.

Let us assume the definition of Art. 194 as the basis of the operation. Let F be the fixed point and OY the fixed line. Let P be the generating point in any position of its path.

Draw FO $\perp$ to OY, and take OY and OX as co-ordinate axes. Draw PL $\parallel$ to OY, PD $\perp$ to OY, and join P and F. Let OF $= p$.

By definition $\dfrac{FP}{DP} = e = $ a constant.

From triangle FPL, $FP^2 = FL^2 + PL^2$; . . . (1) but $FL^2 = (OL - OF)^2 = (x - p)^2$, $LP^2 = y^2$; and $FP^2 = e^2DP^2 = e^2x^2$.

These values in (1) give

$$e^2 x^2 = (x - p)^2 + y^2$$

or, after reduction,

$$y^2 + (1 - e^2) x^2 - 2\,px + p^2 = 0 \ . \ . \ . \ (2)$$

for the required equation.

Cor. Comparing (2) with (1), Art. 138, we find

$$a = 1,\ b = 0,\ \text{and } c = (1 - e^2),$$

hence $\qquad b^2 - 4\,ac = -4\,(1 - e^2) = 4\,(e^2 - 1) \ . \ . \ . \ (3)$

Case 1. *The fixed point not on the fixed line;* i.e., p not zero.

If $e < 1$, $b^2 < 4\,ac$; hence equation (2) is the equation of an ellipse.

If $e = 1$, $b^2 = 4\,ac$; hence equation (2) is the equation of a parabola.

If $e > 1$, $b^2 > 4\,ac$; hence equation (2) is the equation of an hyperbola.

Case 2. *The fixed point is on the fixed line,* i.e., $p = 0$.

In this case (2) becomes

$$y^2 + (1 - e^2) x^2 = 0 \ . \ . \ . \ (4)$$

If $e < 1$, equation (4) represents two imaginary lines intersecting at origin.

If $e = 1$, equation (4) represents one straight line (the X-axis).

If $e > 1$, equation (4) represents two straight lines intersecting at the origin.

Hence, *equation (2) represents the conics or one of their limiting cases.*

GENERAL EXAMPLES.

1. Find the point of intersection of the lines

$$x = 2\,z + 1 \atop y = 3\,z + 2 \Big\}, \quad x = z + 2 \atop y = 4\,z + 1 \Big\}$$

and the cosine of the angle between them.

$$Ans. \quad (3,\,5,\,1)\,; \quad \cos\varphi = \frac{5}{\sqrt{28}}$$

2. Required the equation of the line which passes through $(1, -2, 3)$ and is parallel to

$$x = 2\,z + 1 \atop y = 2 - z \Big\}. \qquad Ans. \quad x = 2\,z - 5 \atop y = -z + 1 \Big\}.$$

3. What is the angle between the lines

$$x + z = 2 \atop y = z - 1 \Big\}, \quad x - 2\,z + 3 = 0 \atop y - z = 2 \Big\}\,?$$

$$Ans. \quad \varphi = 90°.$$

4. What is the distance of the point $(-3, 2, -1)$ from the line

$$x + 3\,z + 1 = 0 \atop y = 4\,z + 3 \Big\}\,?$$

5. A line makes equal angles with the co-ordinate axes; required the angles which it makes with the co-ordinate planes.

6. The equation of a surface is $x^2 + y^2 + z^2 - 2\,x - 4\,y - 6\,z = 2$; what does the equation become when the surface is referred to a parallel system of axes, the origin being at $(1, 2, 3)$? $\qquad Ans.\ x^2 + y^2 + z^2 = 16.$

7. Given the line $\dfrac{x + 2\,z = 2}{y - z = 1}\Big\}$, required the projection of the line on XY and the point in which the line pierces the co-ordinate planes. $\qquad Ans.$ in part, $2\,y + x = 4.$

8. Required the distance cut off on the Z and Y axes by the projections of the line $\dfrac{x + 2\,y = 4}{z + 2\,x = 2}\Big\}$ on YZ.

$$Ans. \quad \begin{aligned} z &= -6 \\ y &= \frac{3}{2} \end{aligned}.$$

9. How are the following pair of lines related?

$$x = 2\,z + 2 \atop y = -\,z - 4 \Bigg\} \quad x = 2\,z - 1 \atop y = -\,z + 2 \Bigg\}$$

10. What are the equations of the line which passes through the origin and the point of intersection of the lines

$$x = 2\,z + 1 \atop y = 3\,z + 2 \Bigg\}, \quad x = z + 2 \atop y = 4\,z + 1 \Bigg\}? \qquad Ans. \quad x = 3\,z \atop y = 5\,z \Bigg\}.$$

11. What is the distance of the point (3, 2, − 4) from the origin? What angle does this line make with its projection on XY?

12. A straight line makes an angle of 60° with the X-axis and an angle of 45° with the Y-axis; what angle does it make with the Z-axis? *Ans.* 60°.

13. What are the cosines of the angles which the line

$$x = 3\,z - 1 \atop y = -\,z + 2 \Bigg\}$$ makes with the co-ordinate axes?

14. A line passes through the point (1, 2, 3) and makes angles with X, Y, Z whose cosines are $\dfrac{\sqrt{2}}{2}, \dfrac{1}{2}, \dfrac{1}{2}$, respectively; required

(*a*) the equation of the line,

(*b*) the equation of the plane ⊥ to the line at the point,

(*c*) to show that the projections of the line are ⊥ to the traces of the plane.

15. The directional cosines of two lines are $\dfrac{2}{3}, \dfrac{1}{3}, \dfrac{2}{3}$ and $\dfrac{\sqrt{2}}{2}, \dfrac{1}{2}, \dfrac{1}{2}$. What is the cosine of the angle which they make with each other?

$$Ans. \quad \text{Cos } \varphi = \frac{3 + 2\sqrt{2}}{6}.$$

16. The projecting planes of a line are $x = 3\,z - 1$ and $x = 2\,y + 2$. What is the equation of the plane which projects the line on YZ? *Ans.* $3\,z - 2\,y = 3$.

17. The projections of a line on XZ and YZ each form with the Z-axis an angle of 45°; required the equation of the line which passes through (2, 1, 4) parallel to the line.

$$\textit{Ans.} \quad \left. \begin{array}{l} x - 2 = z - 4 \\ y - 1 = z - 4 \end{array} \right\}, \quad \text{or} \quad \left. \begin{array}{l} x = z - 2 \\ y = z - 3 \end{array} \right\}.$$

18. Find the equation of the line which contains the point $(3, 2, 1)$ and meets the line $\left. \begin{array}{l} x = 2z - 1 \\ y = z - 3 \end{array} \right\}$ at right angles.

19. Given the lines $\left. \begin{array}{l} x = sz + 2 \\ y = 3z - 1 \end{array} \right\}$ and $\left. \begin{array}{l} x = 2z + 2 \\ y = 3z + 1 \end{array} \right\}$; required

(*a*) the value of s in order that the lines may be parallel;

(*b*) the value of s in order that the lines may be perpendicular;

(*c*) the value of s in order that the lines may intersect.

20. The directional cosines of a line are $\dfrac{2}{3}, \dfrac{1}{3}, \dfrac{2}{3}$; required the sines of the angles which the line makes with the co-ordinate planes.

21. Find the equations of the line which passes through the origin and is perpendicular to the two lines $\left. \begin{array}{l} x = 3z + 5 \\ y = 5z + 3 \end{array} \right\}$ and $\left. \begin{array}{l} x = z + 1 \\ y = 2z \end{array} \right\}$. $\textit{Ans.} \quad \left. \begin{array}{l} x = 3z \\ y = -2z \end{array} \right\}.$

22. Find the angle included between the two planes $Ax + By + Cz = D$ and $A'x + B'y + C'z = D'$.

$$\textit{Ans.} \quad \cos^{-1} \frac{AA' + BB' + CC'}{\sqrt{A^2 + B^2 + C^2} \; \sqrt{A'^2 + B'^2 + C'^2}}.$$

23. If two planes are parallel show that the coefficients of the variables in their equations are proportional.

24. Find the condition for perpendicularity of the two planes given in Example 22.

$$\textit{Ans.} \quad AA' + BB' + CC' = 0.$$

CHAPTER IV.

A DISCUSSION OF THE SURFACES OF THE SECOND ORDER.

By A. L. Nelson, M.A., Professor of Mathematics in Washington and Lee University, Va.

EVERY equation involving three variables represents a surface. If the equation be of the first degree the surface will be a plane. If the equation be of a higher degree the surface will be curved. It is proposed in this chapter to determine the nature of the surfaces represented by equations of the second degree involving three variables. The most general form of the equation of the second degree is $Ax^2 + By^2 + Cz^2 + Dxy + Exz + Fyz + Gx + Hy + Iz + K = 0 \ldots (1)$ where the coefficients A, B, C, etc., may be of either sign and of any magnitude. Let us suppose the co-ordinate axes to be rectangular. The form of equation (1) may be simplified by a transformation of axes. Let us turn the axes without changing the origin.

The formulæ of transformation are (Art. 190)

$$x = x' \cos \alpha' + y' \cos \alpha'' + z' \cos \alpha'''$$
$$y = x' \cos \beta' + y' \cos \beta'' + z' \cos \beta'''$$
$$z = x' \cos \gamma' + y' \cos \gamma'' + z' \cos \gamma'''$$

Substituting these values, equation (1) becomes

$$A'x'^2 + B'y'^2 + C'z'^2 + D'x'y' + E'x'z' + F'y'z' + G'x' + H'y' + I'z' + K = 0 \ldots (2).$$

Since the original axes were supposed rectangular the nine angles α', β', γ' etc., are connected by the three relations

$$\cos^2 \alpha' + \cos^2 \beta' + \cos^2 \gamma' = 1.$$
$$\cos^2 \alpha'' + \cos^2 \beta'' + \cos^2 \gamma'' = 1.$$
$$\cos^2 \alpha''' + \cos^2 \beta''' + \cos^2 \gamma''' = 1.$$

If we take the new axes also rectangular, which is desirable, the nine angles will be connected by the three additional relations

$$\cos \alpha' \cos \alpha'' + \cos \beta' \cos \beta'' + \cos \gamma' \cos \gamma'' = 0.$$
$$\cos \alpha' \cos \alpha''' + \cos \beta' \cos \beta''' + \cos \gamma' \cos \gamma''' = 0.$$
$$\cos \alpha'' \cos \alpha''' + \cos \beta'' \cos \beta''' + \cos \gamma'' \cos \gamma''' = 0.$$

This will leave three of the nine angles to be assumed arbitrarily. Let us give to them such values as to render the coefficients D', E', and F' each equal to zero in equation (2).

The general equation will thus be reduced to the form

$$A'x'^2 + B'y'^2 + C'z'^2 + G'x' + H'y' + I'z' + K = 0,$$

or, omitting accents,

$$Ax^2 + By^2 + Cz^2 + Gx + Hy + Iz + K = 0 \ldots (3)$$

In order to make a further reduction in the form of the equation let us endeavor to move the origin without changing the direction of the axes. The formulæ of transformation will be (Art. 189)

$$x = a + x', \; y = b + y', \; z = c + z'.$$

Equation (3) will become

$$A (a + x')^2 + B (b + y')^2 + C (c + z')^2 + G (a + x') + H (b + y') + I (c + z') + K = 0.$$

Developing, omitting accents, and placing $Aa^2 + Bb^2 + Cc^2 + Ga + Hb + Ic + K = L$, the equation takes the form

$$Ax^2 + By^2 + Cz^2 + (2 Aa + G) x + (2 Bb + H) y + (2 Cc + I) z + L = 0.$$

In order now to give definite values to the quantities a, b, and c, which were entirely arbitrary, let us assume

$$a = - \frac{G}{2 A}, \; b = - \frac{H}{2 B}, \; c = - \frac{I}{2 C} \text{ or}$$

$$2\,\mathrm{A}a + \mathrm{G} = 0,\ 2\,\mathrm{B}b + \mathrm{H} = 0,\ 2\,\mathrm{C}c + \mathrm{I} = 0 \ \ldots (4)$$

If these values of a, b, and c be finite, the general equation reduces to the form

$$\mathrm{A}x^2 + \mathrm{B}y^2 + \mathrm{C}z^2 + \mathrm{L} = 0 \ \ldots [\mathrm{A}],$$

a form which will be set aside for further examination.

It may be remarked that equations (4) are of the first degree, and will give only one value to each of the quantities a, b, and c, and there is therefore only one position for the new origin.

If, however, either A, B, or C be zero, then a, b, or c will become infinite, and the origin will be removed to an infinite distance. This must be avoided.

Let us suppose $\mathrm{A} = 0$, while B and C are finite. We may then assume $2\,\mathrm{B}b + \mathrm{H} = 0$, and $2\,\mathrm{C}c + \mathrm{I} = 0$, but we cannot assume $2\,\mathrm{A}a + \mathrm{G} = 0$.

Having assumed the values of b and c as indicated, let us assume the entire constant term equal to zero. This will give

$$\mathrm{B}b^2 + \mathrm{C}c^2 + \mathrm{G}a + \mathrm{H}b + \mathrm{I}c + \mathrm{K} = 0,$$
$$\text{or } a = -\frac{\mathrm{B}b^2 + \mathrm{C}c^2 + \mathrm{H}b + \mathrm{I}c + \mathrm{K}}{\mathrm{G}}$$

and the general equation will be reduced to the form

$$\mathrm{B}y^2 + \mathrm{C}z^2 + \mathrm{G}x = 0 \ \ldots (\mathrm{B}),$$

a second form set aside for examination.

We must observe that this last proposed transformation will also fail when $\mathrm{G} = 0$, that is, when the first power of x, as well as the second power of x, is wanting in the general equation.

And without making the second transformation we have a third form for examination, viz.:

$$\mathrm{B}y^2 + \mathrm{C}z^2 + \mathrm{H}y + \mathrm{I}z + \mathrm{K} = 0 \ \ldots (\mathrm{C})$$

Lastly, two of the terms involving the second powers of the variables may be wanting, and the equation (1) then becomes

$$\mathrm{C}z^2 + \mathrm{G}x + \mathrm{H}y + \mathrm{I}z + \mathrm{K} = 0 \ \ldots (\mathrm{D})$$

It is apparent, therefore, that every equation of the second degree involving three variables can be reduced to one or another of the four forms

$$Ax^2 + By^2 + Cz^2 + L = 0 \ . \ . \ . \ (A)$$
$$By^2 + Cz^2 + Gx = 0 \ . \ . \ . \ (B)$$
$$By^2 + Cz^2 + Hy + Iz + K = 0 \ . \ . \ . \ (C)$$
$$Cz^2 + Gx + Hy + Iz + K = 0 \ . \ . \ . \ (D)$$

We will examine each of these forms in order, beginning with the first form :

$$Ax^2 + By^2 + Cz^2 + L = 0 \ . \ . \ . \ (A)$$

This equation admits of several varieties of form according to the signs of the coefficients.

1. A, B, and C positive, and L negative in the first member.

2. A, B, C, and L positive.

3. Two of the coefficients as A and B positive, C and L negative.

4. Two of the coefficients as A and B positive, C negative, and L positive.

No other cases will occur.

CASE 1.　　　$Ax^2 + By^2 + Cz^2 = L,$

in which form all of the coefficients are positive.

In order to determine the nature of the surface represented by this equation, let it be intersected by systems of planes parallel respectively to the co-ordinate planes. The equations of these intersecting planes will be $x = a$, $y = b$, $z = c$. Combining the equations of these planes with that of the surface, we find the equations of the projections on the co-ordinate planes of the curves of intersection.

When $x = a$,　　$By^2 + Cz^2 = L - Aa^2$　　an ellipse.
　"　$y = b$,　　$Ax^2 + Cz^2 = L - Bb^2$　　an ellipse.
　"　$z = c$,　　$Ax^2 + By^2 = L - Cc^2$　　an ellipse.

Thus we see that the sections parallel to each of the co-ordinate planes are ellipses.

The section made by the plane $x = a$ is real when $L - Aa^2 > 0$ or $a < \pm \sqrt{\dfrac{L}{A}}$, and imaginary in the contrary case.

The section made by the plane $y = b$ is real when $b < \pm \sqrt{\dfrac{L}{B}}$, and imaginary when $b > \pm \sqrt{\dfrac{L}{B}}$.

The section made by the plane $z = c$ is real when $c < \pm \sqrt{\dfrac{L}{C}}$, and imaginary when $c > \pm \sqrt{\dfrac{L}{C}}$.

Thus we see that the surface is enclosed within a rectangular parallelopiped whose dimensions are

$$2\sqrt{\frac{L}{A}},\ 2\sqrt{\frac{L}{B}}\ \text{and}\ 2\sqrt{\frac{L}{C}}.$$

When $a = \pm \sqrt{\dfrac{L}{A}}$ or $b = \pm \sqrt{\dfrac{L}{B}}$ or $c = \pm \sqrt{\dfrac{L}{C}}$ the sections become points.

When $a = 0$, $b = 0$, and $c = 0$, we find the sections made by the co-ordinate planes to be

$$By^2 + Cz^2 = L.$$
$$Ax^2 + Cz^2 = L.$$
$$Ax^2 + By^2 = L.$$

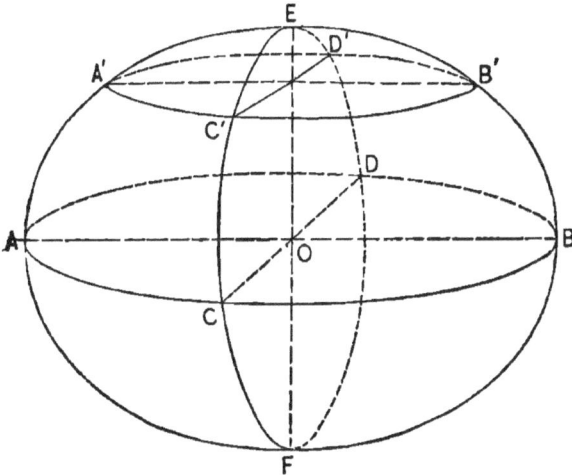

FIG. A.

These are called the *principal* sections of the surface. The principal sections are larger ellipses than the sections parallel to them, as is indicated by the magnitude of the absolute term.

The surface is called the *Ellipsoid*.

It may be generated by the motion of an ellipse of variable dimensions whose centre remains on a fixed line, and whose plane remains always perpendicular to that line, and whose semi-axes are the ordinates of two ellipses which have the same transverse axis, but unequal conjugate axes placed at right angles to each other.   The axes of the principal sections are called the axes of the ellipsoid.

If we represent the semi-axes of the ellipsoid by a, b, and c, we shall have

$$a = \sqrt{\frac{L}{A}}, \; b = \sqrt{\frac{L}{B}}, \; c = \sqrt{\frac{L}{C}}$$

and the equation of the surface

$$Ax^2 + By^2 + Cz^2 = L \text{ becomes}$$

$$\frac{x^2}{a^2} + \frac{y^2}{b^2} + \frac{z^2}{c^2} = 1, \text{ or}$$

$$b^2c^2x^2 + a^2c^2y^2 + a^2b^2z^2 = a^2b^2c^2.$$

These are the forms in which the equation of the ellipsoid is usually given.

If we suppose $B = A$, then $b = a$, and the equation becomes

$$\frac{x^2 + y^2}{a^2} + \frac{z^2}{c^2} = 1,$$

and the surface is the *Ellipsoid of Revolution* about the axis of Z.

If $A = B = C$, then $a = b = c$, and the equation becomes $x^2 + y^2 + z^2 = a^2$, and the surface is a *sphere*.

If $L = 0$, the axes $2\sqrt{\frac{L}{A}}, 2\sqrt{\frac{L}{B}}, 2\sqrt{\frac{L}{C}}$

reduce to zero, and the ellipsoid becomes a *point*.

CASE 2. If L be negative in the second member, the equation $Ax^2 + By^2 + Cz^2 = -L$ will represent an imaginary surface, and there will be no geometrical locus.

Hence the varieties of the *ellipsoid* are

(1) The ellipsoid proper with three unequal axes.

(2) The ellipsoid of revolution with two equal axes.

(3) The sphere.

(4) The point.

(5) The imaginary surface.

CASE 3. In this case the equation takes the form

$$Ax^2 + By^2 - Cz^2 = L,$$

in which A, B, C, and L are essentially positive.

Cutting the surface by planes as before, the sections will be, when $x = a$, $By^2 - Cz^2 = L - Aa^2$, a hyperbola, having its transverse axis parallel to the Y-axis when $a < \pm\sqrt{\dfrac{L}{A}}$, but parallel to the Z-axis when $a > \pm\sqrt{\dfrac{L}{A}}$. And when $a = \pm$ $\sqrt{\dfrac{L}{A}}$, the intersection becomes two straight lines whose projections on the plane of YZ pass through the origin.

When $y = b$, $Ax^2 - Cz^2 = L - Bb^2$, a hyperbola, with similar conditions as above.

When $z = c$, $Ax^2 + By^2 = L + Cc^2$, an ellipse real for all values of c.

Since the elliptical sections are all real, the surface is continuous, or it consists of a single sheet.

The principal sections are found by making successively

$a = 0$, which gives $By^2 - Cz^2 = L$, a hyperbola.

$b = 0$,   "   " $Ax^2 - Cz^2 = L$,   "

$c = 0$,   "   " $Ax^2 + By^2 = L$, an ellipse.

The surface is called the *elliptical hyperboloid of one sheet.* The equation may be reduced to the form

$$\frac{x^2}{a^2} + \frac{y^2}{b^2} - \frac{z^2}{c^2} = 1.$$

This surface may be generated by an ellipse of variable dimensions whose centre remains constantly on the Z-axis, and whose plane is perpendicular to that axis, and whose semi-axes are the ordinates of two hyperbolas having the same conjugate axis coinciding with the Z-axis, but having different transverse axes placed at right angles to each other.

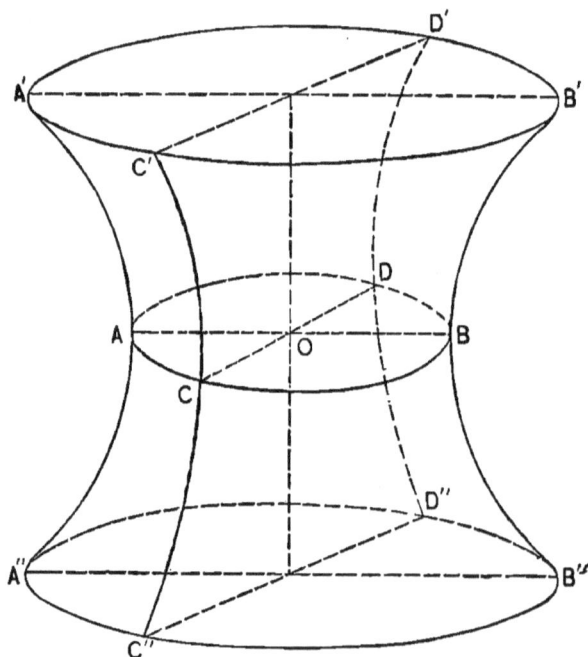

FIG. B.

If we suppose A = B, then will $a = b$, and the equation of the surface becomes

$$\frac{x^2 + y^2}{a^2} - \frac{z^2}{c^2} = 1,$$

the hyperboloid of revolution of one sheet.

If A = B = C, we have $x^2 + y^2 - z^2 = a^2$, the equilateral hyperboloid of revolution of one sheet.

If L = 0, the equation represents a right cone having an elliptical base; and if A = B this base becomes a circle.

Hence we have the following varieties of the *hyperboloid of one sheet.*

1. The hyperboloid proper, with three unequal axes.
2. The hyperboloid of revolution.
3. The equilateral hyperboloid of revolution.
4. The cone.

CASE 4. $Ax^2 + By^2 - Cz^2 = - L$, where A, B, C, and L are essentially positive.

Intersecting the surface as before we have, when $x = a$, $By^2 - Cz^2 = - L - Aa^2$, a hyperbola having its transverse axis parallel to the axis of Z.

When $y = b$, $Ax^2 - Cz^2 = - L - Bb^2$. A hyperbola having its transverse axis parallel to the axis of Z.

When $z = c$, $Ax^2 + By^2 = - L + Cc^2$, an ellipse real when $c > \pm \sqrt{\dfrac{L}{C}}$, and imaginary when $c < \pm \sqrt{\dfrac{L}{C}}$. Since the sections between the limits $z = \pm \sqrt{\dfrac{L}{C}}$ are imaginary, but real beyond those limits, it follows that there are two distinct sheets entirely separated from each other.

The surface is called *the hyperboloid of two sheets.*

The principal sections are found by making successively

$a = 0$, which gives $By^2 - Cz^2 = - L$, a hyperbola with its transverse axis coinciding with the Z-axis.

$b = 0$, which gives $Ax^2 - Cz^2 = - L$, a hyperbola with its transverse axis coinciding with the Z-axis.

$c = 0$, which gives $Ax^2 + By^2 = - L$, an imaginary ellipse.

The semi-axes of the first section are $\sqrt{\dfrac{L}{B}}$ and $\sqrt{\dfrac{L}{C}}$.

Those of the second section are $\sqrt{\dfrac{L}{A}}$ and $\sqrt{\dfrac{L}{C}}$. And those of the imaginary section are $\sqrt{\dfrac{L}{A}(- 1)}$ and $\sqrt{\dfrac{L}{B}(- 1)}$.

The distances $2\sqrt{\dfrac{L}{A}}, 2\sqrt{\dfrac{L}{B}}$, and $2\sqrt{\dfrac{L}{C}}$ are called the axes

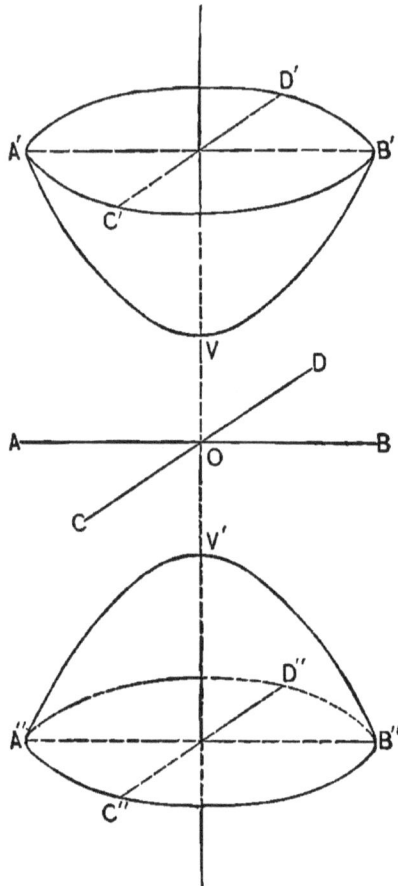

Fig. C.

of the surface. Representing the semi-axes by a, b, and c, the equation of the surface may be reduced to the form

$$-\frac{x^2}{a^2} + \frac{y^2}{b^2} - \frac{z^2}{c^2} = -1.$$

If we suppose A = B, then $a = b$, and the equation reduces to

$$\frac{x^2 + y^2}{a^2} - \frac{z^2}{c^2} = -1,$$

the hyperboloid of revolution of two sheets.

If $A = B = C$, the equation becomes
$$x^2 + y^2 - z^2 = - a^2,$$
which represents the surface generated by the revolution of an equilateral hyperbola about its transverse axis.

Finally, if $L = 0$, the surface becomes a cone having an elliptical base, and the base becomes a circle when $A = B$.

We have, therefore, the following varieties of the *hyperboloid of two sheets:*
1. The hyperboloid proper having three unequal axes.
2. The hyperboloid of revolution.
3. The equilateral hyperboloid of revolution.
4. The cone.

We will now examine the second form,
$$By^2 + Cz^2 + G x = 0 \ \ldots \ \text{(B)}$$
Three cases apparently different present themselves for examination.

(1). B and C positive and G negative in the first member.

(2). B, C, and G positive.

(3). B positive and C and G negative.

CASE 1. The equation may be written
$$By^2 + Cz^2 = Gx$$
in which B, C, and G are essentially positive.

Let the surface be intersected as usual by planes parallel respectively to the co-ordinate planes.

When $x = a$, $By^2 + Cz^2 = Ga$, an ellipse real when $a > 0$, and imaginary when $a < 0$.

When $y = b$, $Cz^2 = Gx - Bb^2$, a parabola with its axis parallel to the axis of X.

When $z = c$, $By^2 = Gx - Cc^2$ a parabola with its axis parallel to the axis of X.

The principal sections are found by making $a = 0, b = 0$, and $c = 0$.

When $a = 0$, $By^2 + Cz^2 = 0$, a point, the origin.

When $b = 0$, $Cz^2 = Gx$, a parabola with its vertex at the origin.

When $c, = 0, By^2 = Gx$, a parabola with vertex at the origin.
Since every positive value of x gives a real section, and
every negative value of x an imaginary section, the surface
consists of a single sheet extending indefinitely and contin-
uously in the direction of positive abscissas, but having no
points in the opposite direction from the origin.

The surface is called *the elliptical paraboloid*. It may be
generated by the motion of an ellipse of variable dimensions
whose centre remains constantly on the same straight line,
and whose plane continues perpendicular to that line, and
whose semi-axes are the ordinates of two parabolas having a
common axis and the same vertex, but different parameters
placed with their planes perpendicular to each other.

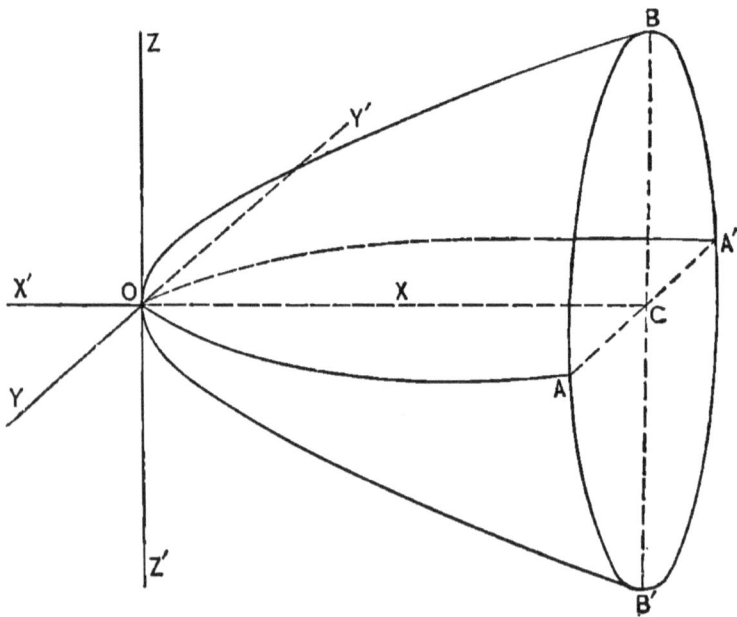

FIG. D.

CASE 2. If we suppose G to be positive in the first member
so that the equation will take the form

$$By^2 + Cz^2 = - Gx,$$

the sections perpendicular to the X-axis will become imaginary when $x > 0$, and real when $x < 0$.

In other respects the results are similar to those deduced in case 1.

Thus the equation will represent a surface of the same form as in case 1, but turned in the opposite direction from the co-ordinate plane of YZ.

If $B = C$, the surface becomes the paraboloid of revolution.

CASE 3. $By^2 - Cz^2 = Gx.$

Intersect the surface by planes as before.

When $x = a$, $By^2 - Cz^2 = Ga$, a hyperbola with transverse axis in the direction of the Y-axis when $a > 0$, and in the direction of the Z-axis when $a < 0$.

When $y = b$, $Cz^2 = -Gx + Bb^2$, a parabola having its axis in the direction of the X-axis and extending to the left.

When $z = c$, $By^2 = Gx + Cc^2$, a parabola having its axis in the direction of the X-axis and extending to the right.

Since every value of x, either positive or negative, gives a real section, the surface consists of a single sheet extending indefinitely to the right and left of the plane of YZ. This surface is called the *Hyperbolic Paraboloid*. To find its principal sections make x, y, and z alternately equal to zero.

When $x = 0$, $By^2 = Cz^2$, two straight lines.

When $y = 0$, $Cz^2 = -Gx$, a parabola with axis to the left.

When $z = 0$, $By^2 = Gx$, a parabola with axis to the right.

The hyperbolic paraboloid admits of no variety.

Now taking up form (C), $By^2 + Cz^2 + Hy + Iz + K = 0$, we see that it is the equation of a cylinder whose elements are perpendicular to the plane of YZ, and whose base in the plane of YZ will be an ellipse or hyperbola according to the signs of B and C.

The fourth form (D), $Cz^2 + Gx + Hy + Iz + K = 0$ represents a cylinder having its bases in the planes XZ and YZ

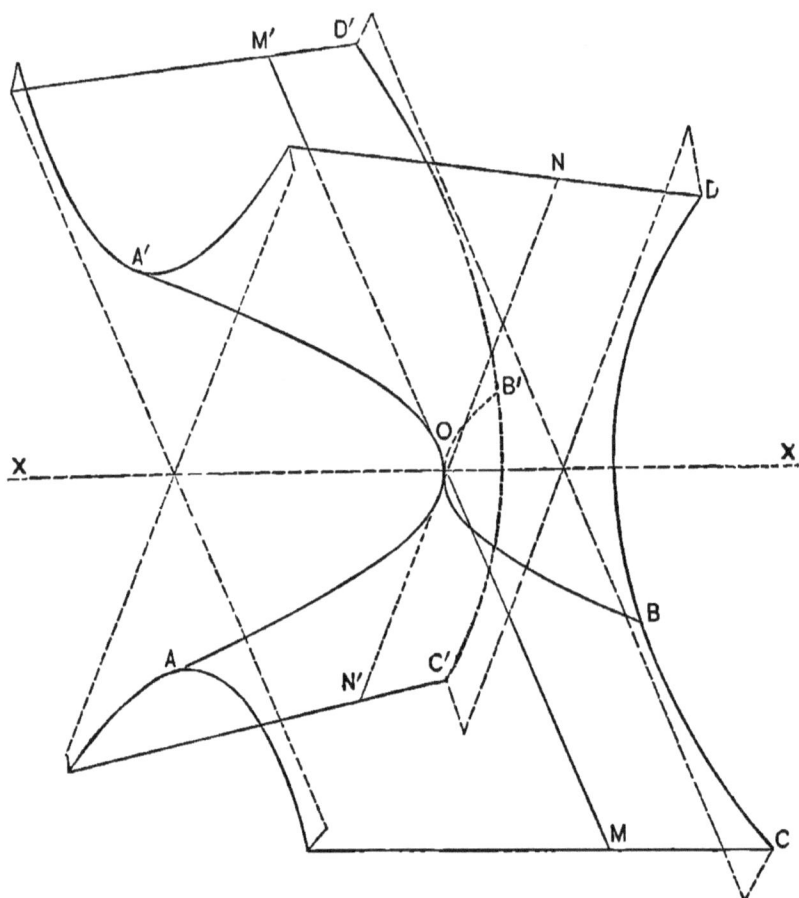

FIG. E.

parabolas, and having its right-lined elements parallel to the plane XY and to each other, but oblique to the axes of X and Y.

The preceding discussion shows that every equation of the second degree between three variables represents one or another of the following surfaces:

1. The ellipsoid with its varieties, viz.; the ellipsoid proper, the ellipsoid of revolution, the sphere, the point, and the imaginary surface.

2. The hyperboloid of one or two sheets, with their varieties, viz.: the hyperboloid proper of one or two sheets, the hyperboloid of revolution of one or two sheets, the equilateral hyperboloid of revolution of one or two sheets, the cone with an elliptical or circular base.

3. The paraboloid, either elliptical or hyperbolic, with the variety, the paraboloid of revolution.

4. The cylinder, having its base either an ellipse, hyperbola, or parabola.

Surfaces of Revolution. — The general equation of surfaces of revolution may be deduced by a direct method, as follows :

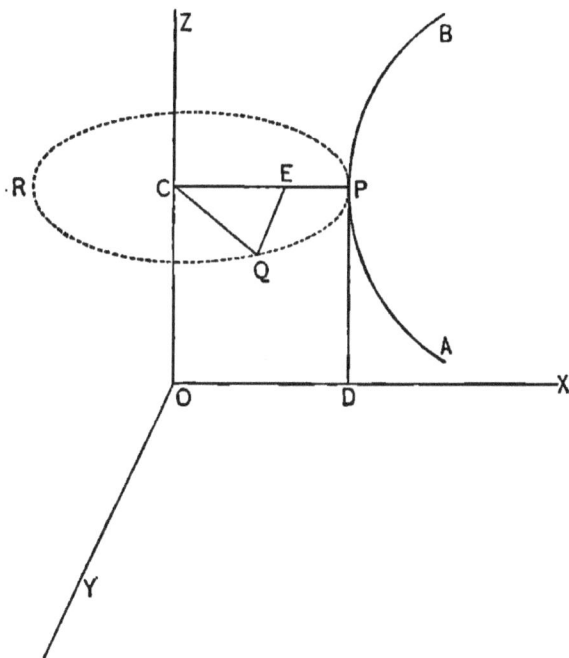

FIG. F.

Let the Z-axis be the axis of revolution, and let the equation of AB, the generating curve in the plane of XZ, be $x^2 = fz$.

Let P be the point in this curve which generates the circle

PQR, and let r be the radius of the circle. We will have $r^2 = x^2 + y^2$.

The value of r^2 may also be expressed in terms of z from the equation of the generatrix in the plane of XZ as follows:

$$r^2 = \overline{CP}^2 = \overline{OD}^2 = fz.$$

Equating these two values of r we have

$$x^2 + y^2 = fz$$

as the general equation of surfaces of revolution.

It will be observed that the second value of r^2 is the value of x^2 in the equation of the generatrix. Hence, to find the equation of the surface of revolution we have only to substitute $x^2 + y^2$ of the surface for x^2 in the generatrix.

Surface of a Sphere. — Equation of generatrix $x^2 + z^2 = R^2$. Hence the equation of the surface of the sphere is

$$x^2 + y^2 + z^2 = R^2.$$

Ellipsoid of Revolution. —

Generatrix $\dfrac{x^2}{a^2} + \dfrac{z^2}{c^2} = 1.$

Surface $\dfrac{x^2 + y^2}{a^2} + \dfrac{z^2}{c^2} = 1.$

Similarly, the equation of the hyperboloid of revolution is

$$\frac{x^2 + y^2}{a^2} - \frac{z^2}{c^2} = 1.$$

Paraboloid of Revolution. —

$x^2 = 4pz$, the generatrix.

$x^2 + y^2 = 4pz$, the surface of revolution.

Cone of revolution. $z = mx + \beta$ the generatrix,

or $x = \dfrac{z - \beta}{m}$, $x^2 = \dfrac{(z - \beta)^2}{m^2}$.

Hence $x^2 + y^2 = \dfrac{(z - \beta)^2}{m^2}$

or $m^2(x^2 + y^2) = (z - \beta)^2$.

EXAMPLES.

1. What is the locus in space of $4\,x^2 + 9\,y^2 = 36$? Of $9\,z^2 - 16\,y^2 = 144$? Of $x^2 + y^2 = r^2$? Of $y^2 + z^2 = r^2$? Of $y^2 + 8\,x = 0$?

2. Determine the nature of the surfaces $x^2 + y^2 + 4\,z^2 = 25$, $7\,(x^2 + y^2) - 4\,z^2 = 79$.

3. Find the equation of the surface of revolution about the axis of Z whose generatrix is $z = 3\,x + 5$.

4. Find the equation of the cone of revolution whose intersection with the plane of XY is $x^2 + y^2 = 9$, and whose vertex is $(0, 0, 5.)$

5. Determine the surfaces represented by

$$x^2 + 4\,y^2 + 9\,z^2 = 36.$$
$$x^2 + 4\,y^2 - 9\,z^2 = 36.$$
$$x^2 + 4\,y^2 = 9\,z^2 - 36.$$
$$4\,y^2 + 9\,z^2 = 36\,x.$$
$$4\,y^2 - 9\,z^2 = 36\,x.$$
$$9\,z^2 - 4\,y^2 = 36\,x.$$